WHO WILL MIND THE BABY?

Who Will Mind the Baby? offers geographic explorations of child care at a variety of geographic scales within the context of US and Canadian case studies. The book is organized around two themes: the government's role in child care provision, and the child care arrangements and everyday geographies of working mothers.

Child care provision is geographically uneven in North America, reflecting the haphazard manner in which child care provision has evolved. Neither the US nor Canada has a national child care policy – in stark contrast to most European countries, where child care is a national concern. The book explores the sociospatial implications of public policies in the US and Canada relating to the provision, funding, and regulation of child care, employing case studies pitched at a variety of scales: national, regional, rural, metropolitan, and local levels.

The book also discusses how mothers' daily routines link the varied geographies of child care, home and paid work – geographies that invariably do not overlap as child care facilities are often located away from the home and workplace. Using everyday geographies of working mothers, the authors look at journeys to child care, reasons for selecting particular child care arrangements, and the coping strategies of working mothers. Finally the book places child care in international perspective and considers future research on child care provision.

Kim England is Associate Professor of Geography, Scarborough College, University of Toronto.

INTERNATIONAL STUDIES OF WOMEN AND PLACE

Edited by Janet Momsen, *University of California at Davis*
and Janice Monk, *University of Arizona*

The Routledge series of *International Studies of Women and Place* describes the diversity and complexity of women's experience around the world, working across different geographies to explore the processes which underlie the construction of gender and the life-worlds of women.

Other titles in this series:

FULL CIRCLES
Geographies of women over the life course
Edited by Cindi Katz and Janice Monk

'VIVA'
Women and popular protest in Latin America
Edited by Sarah A. Radcliffe and Sallie Westwood

DIFFERENT PLACES, DIFFERENT VOICES
Gender and development in Africa, Asia and Latin America
Edited by Janet H. Momsen and Vivian Kinnard

SERVICING THE MIDDLE CLASSES
Class, gender and waged domestic labour in contemporary Britain
Nicky Gregson and Michelle Lowe

WOMEN'S VOICES FROM THE RAINFOREST
Janet Gabriel Townsend

GENDER, WORK, AND SPACE
Susan Hanson and Geraldine Pratt

WOMEN AND THE ISRAELI OCCUPATION
Edited by Tamar Mayer

FEMINISM/POSTMODERNISM/DEVELOPMENT
Edited by Marianne H. Marchand and Jane L. Parpart

WOMEN OF THE EUROPEAN UNION
The Politics of Work and Daily Life
Edited by Janice Monk and Maria Dolors García Ramon

WHO WILL MIND THE BABY?

Geographies of child care and working mothers

Edited by Kim England

London and New York

First published 1996
by Routledge
11 New Fetter Lane, London EC4P 4EE

Simultaneously published in the USA and Canada
by Routledge
29 West 35th Street, New York, NY 10001

Typeset in Baskerville by LaserScript, Mitcham, Surrey
Printed and bound in Great Britain by
Clays Ltd, St. Ives PLC

British Library Cataloguing in Publication Data
A catalogue record for this book is available from the British
Library

Library of Congress Cataloging in Publication Data
Who will mind the baby?: geographies of child care and
working mothers
/edited by Kim England.
p. cm. – (International studies of women and place)
Includes bibliographical references and index.
1. Child care – Government policy. 2. Child care services –
United States. 3. Child care services – Canada. I. England, Kim.
II. Series.
HQ778.5.W48   1996
362.7′12′0973 – dc20   95-52129

ISBN 0–415–11740–2 (hbk)
ISBN 0–415–11741–0 (pbk)

CONTENTS

ILLUSTRATIONS

FIGURES

TABLES

CONTRIBUTORS

David E. Bloom is Professor at the Department of Economics at Columbia University.

Susan R. Brooker-Gross is Associate Professor at the Department of Geography at the Virginia Polytechnic Institute and State University.

Ellen K. Cromley is Associate Professor at the Department of Geography at the University of Connecticut.

Isabel Dyck is Associate Professor at the School of Rehabilitation Sciences, Faculty of Medicine at the University of British Columbia.

Kim England is Associate Professor at the Division of Social Sciences at Scarborough College, University of Toronto.

Ruth Fincher is Reader at the Department of Geography at the University of Melbourne.

Holly J. Myers-Jones is Visiting Assistant Professor at the Department of Geography and Planning at Toledo State University.

Ian Skelton is Associate Professor at the Department of City Planning at the University of Manitoba.

Todd P. Steen is Associate Professor at the Department of Economics and Business Administration, Hope College, Michigan.

Marie Truelove is Professor at the Department of Geography at Ryerson Polytechnical University, Toronto.

ACKNOWLEDGEMENTS

This book has been so much longer in the making than I could have ever imagined. Along the way a number of people have provided support and assistance of various sorts for which I am most appreciative. In particular I thank Mike Bunce, Carl Drouin, Muriel England, Meric Gertler, Audrey Glasbergen, Gustav Gunterson, Lille Havfrue, Bev Kueper, Kathy Morihovitis, Kelly Oliver, Linda Peake, Ted Relph and Sue Ruddick. In addition to their encouragement and guidance since the beginning, Janet Momsen and Jan Monk provided helpful comments on the manuscript. Tristan Palmer and Matthew Smith at Routledge offered sound advice, while valiantly and politely keeping me to schedule. Caroline Cautley, Katherine Hodkinson and Sophie Richmond proved to be great editors. I thank the Department of Geography at Miami University, Ohio for granting me permission to let a number of the authors re-work and update papers that had originally appeared in *The Ohio Geographer*. And, of course, thanks to the contributors who endured these re-writes without complaint!

Part 1

INTRODUCTION

1

WHO WILL MIND THE BABY?

Kim England

One of the most significant recent social and economic changes in North America is the explosion in the numbers of mothers in paid employment. In the last twenty-five years women with children aged less than 6 years old have been the fastest growing group of women entering paid employment. Yet the demand for child care continues to outstrip supply, often forcing parents to make less than satisfactory arrangements. This, along with the continued trend toward large numbers of women with young children in the labor force, highlights the pressing need for good quality affordable and accessible child care. However, unlike many European countries, neither the US nor Canada has a national child care policy. Rather than offering a comprehensive system of universally available child care, both countries have policies and pieces of legislation generated and implemented by various levels of government for various purposes at various points in time. This haphazard history and fragmentation compels many commentators to describe child care in Canada and the US as "patchwork" and "piecemeal." This patchwork system means that there are noticeable spatial variations in child care provision, as well as differential access to that child care in terms of social characteristics (whereas there is often less variation in countries with national child care policies).

Child care, then, is an obvious candidate for an analysis of geographic access to service facilities, spatial equity and social justice. Yet as contributors to this volume note, there is a paucity of geographic studies of this sort (important exceptions include Cromley, 1987; Fincher, 1993; Kanaroglou and Rhodes, 1990; Truelove, 1993a). This is curious because of the strong tradition of social justice and welfare geography focusing on who gets what, where and how (Harvey, 1973, 1992; Smith, 1977, 1994). The "where" reflects an understanding that living standards and access vary according to area of residence; and, as Holly Myers-Jones and Susan Brooker-Gross argue in their chapter in this volume, "(s)patial equity questions arise as service provision is compared with proximity to residential neighborhoods, themselves differentiated by family composition, economic status, race and ethnicity" (p. 83). Moreover, Ellen Cromley, Ian

3

Skelton and Marie Truelove (Chapter 7) show that spatial equity questions also arise at the scale of the province/state and local municipality. Clearly, then location – at a variety of scales – is an important component of child care, and, as Truelove argues, "(g)eographers can make clearer the spatial impacts of policies that initially appear to most people to be aspatial, and can analyze how social policies affect well-being" (Chapter 7: 107).

The US and Canadian child care "systems" consist of an assortment of policies and pieces of legislation that are not coordinated into a comprehensive national plan. This "patchwork" system means that there is noticeable variation in terms of the type, accessibility, quality and quantity of child care. For instance, Truelove (Chapter 3) notes that there are significant differences in the provision and regulation of licensed child care across Canada's provinces and territories (also see Friendly, 1994; Goelman, 1992; Mackenzie and Truelove, 1993), and similar patterns are apparent in the states (Kahn and Kamerman, 1987a). At the same time, many child care programs are the result of local initiatives by parent groups, charities, municipalities or employers, and so there is also substantial variation within the states (Kahn and Kamerman, 1987a; Young and Good, 1986), and the provinces and territories (Prentice, 1988; Truelove, 1993a). In this book the theme of spatial variation is picked up by David Bloom and Todd Steen and by Truelove (Chapters 3 and 7); and spatial equity, while implicit in a number of chapters, is explicitly explored at the state level by Cromley in her discussion of Connecticut and at the provincial level in Skelton's case study of Ontario (also see Truelove, 1993a).

At the heart of this collection is a consideration of child care at a variety of scales within the context of Canadian–US case studies. Many studies exploring child care and working mothers have examined the issue from the national perspective; this is especially true of international comparisons such as Lamb *et al.* (1992) and Melhuish and Moss (1991a). Frequently, the oftentimes substantial state/provincial and local variations in implementing national policies and programs are ignored, as are important initiatives at the state/provincial and local level. At the other end of the spatial spectrum there are a series of local case studies by geographers and non-geographers within the academy (for example: Kahn and Kamerman, 1987b; Rose, 1990), as well as by practitioners, activists and policy makers (for example: Action for Children, 1989; Barnhorst and Johnson, 1991).

Dealing with child care *across* a variety of spatial scales is important because disparities and inequities often become more apparent at finer scales. For instance, both the US and Canada have national policies aimed at increasing access to, and easing the costs of child care for low-income families (be this through tax credits, child care subsidies or certain welfare payments). However, contributors to this volume who pitched their analyses at the level of the state (Cromley), province (Skelton) and small town (Myers-Jones and Brooker-Gross) indicate that child care services are either greater in middle-class areas or that children from middle-class families are over-represented in child care centers. These same studies also provide ample evidence of the desperate shortage of child care in rural areas. In addition, Truelove's (Chapter 7) case study of Metropolitan Toronto reveals that access to child care spaces is greater in the central city, and that while the number of suburban child care centers has grown (some as a direct result of local government policy), suburban families have relatively fewer local child care centers (especially non-profit centers) for their children to attend. Of course, not all suburbs are the same, and Truelove notes that different Toronto suburbs have different mixes of non-profit, commercial and municipal child care centers. Ruth Fincher found similar patterns for Melbourne, and, as she notes in this volume, there are "differences in the propensity of [suburban] governments to take up the opportunity for federal funding of local child care" (1991: 157). All these studies add weight to Ferguson's (1991, quoted by Fincher, this volume) plea that forms of child care should "always be assessed in terms of their implications for women of all classes, races, and geographical locations."

That child care services are geographically uneven is apparent in the daily lives of employed parents. Their daily routines link together the geographies of child care, home and paid work, geographies that invariably do not overlap, as child care facilities are often in quite different locations than home or workplace. Mothers are usually responsible for arranging child care and then chauffeuring the child to it (Michelson, 1985, 1988; Rosenbloom, 1988, 1993). Thus, employed mothers of young children face especially complex time–space budgeting problems in their daily routines, and may be faced by spatial constraints such as poor public transportation and limited opportunities for affordable, quality child care at convenient locations. A number of the contributors develop this theme. Myers-Jones and Brooker-Gross, and Truelove (Chapter 7) focus on aspects of the journey to child care, finding that they are short in distance from home and workplace. Isabel Dyck and Kim England look at the more general coping strategies mothers create in order to weave together their various daily responsibilities.

This book brings together research that deals explicitly with the spatial and social implications of child care and the paid employment of mothers in various contexts, and at a variety of scales within Canada and the US. The book is organized around two main themes: the government's role in child care provision and the everyday geographies of working mothers. In this introductory chapter, I explore these two themes and point out how the contributors' chapters fit into these themes. A number of chapters examine public policies related to the provision, funding and regulation of child care in the US and Canada at the national, state/provincial and local levels. Some comparisons are made with child care policies in European countries and Australia. Other chapters examine child care arrangements and the everyday geographies of working mothers in a variety of local settings, exploring, in part, how outcomes vary by parents' social characteristics. These chapters develop a number of themes relating to the everyday geographies of working mothers, including time geography, coping strategies and the importance of locally embedded knowledge.

CHILD CARE POLICY AND PROVISION IN CANADA AND THE US

Child care in the US and Canada is provided in a variety of ways. Marie Truelove (Chapter 3) describes four types of child care: in-home care, unlicensed family child care homes, licensed family child care homes and child care centers. The first two are informal child care, the second two represent the formal sector. About two-thirds of North American (slightly more US than Canadian) preschoolers are in home-based care (this might be the child's home, but increasingly it is the carer's home) and most of the carers in home-based care are relatives of the child. Some in-home care is provided by nannies. Although this book only deals with this type of care in passing (one of the women whom Isabel Dyck interviewed employed a nanny), it seems to be an increasingly important part of the matrix of child care in Canada and the US, especially for dual-career families (Youcha, 1995). Nicky Gregson and Michelle Lowe (1994) examine the revival of waged domestic labor, including nannies, in the UK during the 1980s and early 1990s. They link the increased demand for waged domestic workers to the rise of the service class, and especially dual-career couples, that, in turn, relates to the growth of women's full-time employment in managerial and professional occupations. Certainly, there has been an increase in nanny and domestic placement agencies in large metropolitan areas across the US and Canada, but more often than not these are unregulated, especially in the US (Bakan and Stasiulis, 1995; England and Stiell, forthcoming; Hayes *et al.*, 1990).

The majority of this book deals with formal child care, especially center care. In the last twenty-five years, the number of formal child care spaces has increased rapidly. Ian Skelton traces this growth for Ontario and Truelove (Chapter 7) explores the evolution in Metropolitan Toronto. Both find a rapid increase in spaces since the early 1970s. David Bloom and Todd Steen note that in the US formal child care spaces have also grown in number, and that since the late 1970s there has been a trend toward increased use of formal care (also see O'Connell and Bloom, 1987; Hofferth and Phillips, 1987; Phillips, 1991). Despite this, currently only between 20 and 25 per cent of North American children are in formal child care. Some child care centers in Canada, and a few in the US, are publicly funded and operated, but the majority are private (either non-profit or commercial), receiving minimal, if any public funding and relying on parents' fees (thus child care costs are a major family expense). This is in strict contrast to most European Union and Scandinavian countries, where child care receives extensive public funding (Friendly, 1994; Lamb *et al.*, 1992; Melhuish and Moss, 1991a). In much of its social policy, Canada has adopted more of a universal, publicly funded, "European-style" approach than the US. But, in terms of child care, Canada is more similar to the US than to most European Union countries. There are, however, some key differences between the two countries, especially in terms of who operates the program. In Canada, some provinces offer publicly operated care run by municipalities or school boards (35 per cent of regulated child care spaces in Québec are publicly operated, compared to 7.5 per cent in Ontario and 0.7 per cent in Saskatchewan; it is not available in the other provinces or the territories). The amount of publicly operated child care is negligible in the US.

One of the major differences between Canada and the US is around the issue of providing child care for profit. Practically non-existent in most European Union and Scandinavian countries, about 50 per cent of child care in the US, compared with 30 per cent of Canadian child care centers are commercial/for-profit (Friendly, 1994; Phillips, 1991). The commercial child care sector has expanded more rapidly in the US than Canada, especially during the 1980s (in 1981 less than a quarter of centers in the US were for-profit). In both countries most for-profit centers are run by single owners. However, the US has seen the rise of child care chains (like Kindercare, which, as Friendly (1994) remarks, not only has some centers in Canada, but is also listed on the New York Stock Exchange). Chains are not common in Canada, but estimates indicate that one in six commercial centers in the US are run by chains (Bloom and Steen, this volume; Fincher, this volume; Gormley, 1995; Hofferth and Phillips, 1987; Kamerman and Kahn, 1989; Phillips, 1991). In Canada, as Skelton and Truelove (Chapter 3) indicate, commercial child care *has* grown as federal funds have become scarce, but the growth of commercial child care lags behind the growth of non-profit child care (many provinces legislate that commercial centers can receive minimal or no public monies) (also see Friendly *et al.*, 1987). The for-profit/non-profit debate relates to a slew of issues, but pivotal is the issue of quality. Evidence from Canada suggests that non-profit is of higher quality (Friendly, 1994; Ontario Coalition for Better Day Care, 1987). In the US, evidence is mixed (Kamerman and Kahn, 1989), although recent studies echo Canadian findings, especially when the focus is on child care chains (Gormley, 1995).

Another important difference is maternity and parental leaves. After a long struggle, the US finally passed legislation regarding maternity leave under the 1993 Federal and Medical Leave Act (although some states already had maternity leave legislation). However, as Bloom and Steen point out, it is *unpaid* leave for twelve weeks with a guarantee that the employee can return to a similar (but not necessarily the same) job. Moreover, this legislation is not universal, for instance it only applies to employers with more than fifty employees and is only granted to employees with one year's employment history with the firm. Maternity leave has existed in Canada since the 1970s. Currently, maternity and parental leave is more generous in Canada than in the US, although it still lags behind many European countries. As Truelove (Chapter 3) describes, people on maternity and parental leave receive 57 per cent of their regular earnings (compared, for example, to 90 per cent in France and Sweden) for fifteen weeks. Most of the provinces and territories have their own maternity and parental leave legislation and require employers to hold the jobs of employees on parental leave (also see Friendly, 1994).

Neither the US nor Canadian governments, however, have a national child care policy of publicly funded child care, nor broadly based family benefits (although Québec provides an important exception (see Goelman, 1992; Lero and Kyle, 1991; Rose, 1990), in that it *does* have a family-focused set of policies and programs). In contrast, Scandinavian countries and, to a lesser extent, most European Union countries have elected to make child care a significant national priority, and their strong commitment is apparent in national child care policies complemented by a set of family policies including generous (certainly by North American standards) parental and family leaves, and flexible working hours. These policies afford parents some flexibility regarding combining parenting and paid employment. Child care is relatively cheap and heavily subsidized, and local government usually plays an important role in service delivery (especially in Scandinavia) (Kamerman, 1989; Lamb *et al.*, 1992; Melhuish and Moss, 1991a).

The Canadian and US governments approach child care in what Bronfenbrenner (1992) describes as the "Anglo-Saxon mode." Like the United Kingdom, their philosophy regarding child care is marked by a commitment to individualism and family privacy (Friendly, 1994; Phillips, 1991). In Chapter 10 Ruth Fincher reviews different countries' philosophies toward child care. She cites the US (along with the UK) as an extreme example of a country committed to viewing child care as a private responsibility, as well as holding the opinion that leaving child care provision to the "market" increases individual families' choice (Friendly, 1994; Lamb *et al.*, 1992; Melhuish and Moss, 1991a).

Two further issues set the US and Canada apart from most European Union and Scandinavian countries: first, in Canada and the US child care gets identified as an employment related issue rather than a family issue (which helps explain why neither country has a clearly defined family policy); and second, historically both countries have approached publicly funded child care as social assistance or welfare. First, in the US and Canada the supply of child care gets directly linked to the paid employment of mothers; whereas in France and Sweden, for example, supply is relative to *all* children regardless of their mother's paid employment status and child care tends to be viewed as a community rather than a private issue (Beach, 1992; Bowlby, 1990; Friendly, 1994; Kamerman and Kahn, 1989; Melhuish and Moss, 1991a). That child care is approached as a (women's) employment issue rather than a family issue is brought out in a number of the chapters (Bloom and Steen, Truelove, Fincher and Skelton). Bloom and Steen explore how expanding the "child care industry" would increase the supply of women available for paid employment, especially low-income women and lone mothers (Skelton also touches on this issue in Chapter 5). Moreover, that child care provision is linked to labor market policies explains why so much attention is paid to the role of employers in child care provision. Despite the amount of media attention that employer-sponsored child care receives in Canada and the US, Bloom and Steen remark that it still represents a very small proportion of child care in North America and tends only to be available at some government offices, universities, hospitals and a few large companies. Truelove (Chapter 3 and 7) notes that in Metropolitan Toronto, workplace child care tends to be focused on the downtown core. Some of the first on-site care was offered at workplaces (such as hospitals) that operated around shift-work. Child care provision beyond the conventional working day, whether on-site or not, continues to be notoriously scarce (Friendly, 1994; Gormley, 1995). Basically, the supply of workplace-based child care is very uneven, and, as Fincher remarks, even within the same workplace "'lower status employees' in firms, especially women, are likely to have less access to on-site child care and to flexible schedules that make parenting and employment easier" (p. 155).

However, as Bloom and Steen point out, employers are increasingly offering other forms of support for child care, including non-taxable benefits deducted from salaries, voucher programs for nearby child care centers, and child care referral and counseling services. Resource and referral agencies might be in-house or contracted, they provide information on the quality, cost and availability of child care (Hayes *et al.*, 1990; Gormley, 1995). Bloom and Steen remark that these agencies grew rapidly during the 1980s, and that they are often an important part of a firm's child care strategy. For instance, Myers-Jones and Brooker-Gross note that the university in their study has an in-house resource and referral center that opened in the late 1980s. England (Chapter 8) interviewed employers about their policies (if any) to create "family friendly workplaces," in terms of flexible work schedules and child care benefits and vouchers (also see Beach *et al.*, 1993; Bloom and Trahan, 1987; Cromley, 1987; Kamerman and Kahn, 1989; Lero and Kyle, 1991). Work-related child care, however, raises questions about increased employer control. Employees dependent on workplace child care may find it difficult to leave the company or be involved in union activity because of concerns for their child's well-being. At the same time, on-site or near-site child care also places the burden of child care responsibility on the parent who is employed there (Cohen, 1993; Cromley, 1987; Melhuish and Moss, 1991b). And, as Bloom and Steen, Myers-Jones and Brooker-Gross and Truelove remark in their chapters, one reason many parents favor child care in their neighborhood is that young children do not always cope well with lengthy commutes (also see Bowlby, 1990; Cromley, 1987).

Another issue that sets the US and Canada apart from most European Union and Scandinavian countries is that publicly funded child care has historically been associated with welfare services for, or social assistance to low-income parents.[1] Cromley captures this "welfarist" approach to child care in her discussion of the emergence of the "environment-of-the-child issue" to refer to children "disadvantaged" by their home environment. More generally, Bloom and Steen describe how in the US, federal funds that can be used to subsidize child care filter to the states through a variety of welfare programs not always specifically aimed at child care: for example, Project Head Start (an early childhood education program for low-income children, intended to minimize their future poverty); and two block grants – the Social Services Block Grants (known as Title XX of the 1975 Social Security Act, until 1981) that allow states to allocate federal funds to a variety of social services (including child care), and the Child Care and Development Block Grant (part of the 1990 Omnibus Budget Reconciliation Act) to improve quality and purchase child care for disadvantaged children. On the demand side, child care subsidies are available to welfare-eligible mothers, through, for example, Aid to Families with Dependent Children (AFDC), funded through Title IV-A of the 1975 Social Security Act (Kahn and Kamerman, 1987a; Kamerman and Kahn, 1989; Bloom and Steen, this volume). Truelove (Chapter 3) outlines the Canadian system of funding through the Canada Assistance Plan (CAP), a federal/provincial cost-sharing scheme put in place in 1966 to fund a range of social programs that the provinces and territories implement and regulate. CAP includes a provision for welfare services to fund certain services in order to *prevent* poverty, and a provision for social assistance to *alleviate* poverty faced by families "in need" (Friendly, 1994; Mackenzie and Truelove, 1993; Skelton, this volume).[2]

Recently, there has been growing demand in Canada and the US that extensive reliance on non-parental child care by families in all income brackets necessitates a more comprehensive child care system. Moreover, and as Truelove (Chapter 3: 45) remarks, there is increased recognition that child care "be perceived as an essential and 'mainstream service' not just a welfare-oriented service." However, increased government intervention raises questions about public (or collective) versus private (or individual) responsibility for child care, questions that conflict with the "Anglo-Saxon mode" of individualism and family privacy. At the same time, a number of the contributors (Cromley, Myers-Jones and Brooker-Gross, and Fincher) point to another set of questions around the educational component of child care that has received attention with the increased use of child care by middle-class parents (Friendly, 1994; Lero and Kyle, 1991; Moss and Melhuish, 1991b). Rose suggests that "educated middle-class families place considerable value on the formal system as a means of socialization of children" (1990: 372). These two sets of questions overlap, of course. Some issues that had previously been considered "private," personal problems requiring individual solutions that over time have been redefined as community problems requiring changes in public policy. Education is one example. Elementary and secondary education, and more recently higher education have gradually been recast as public goods that result in advantages for broader society. As yet, child care is not seen in this light (unlike many European Union and Scandinavian countries)(Gormley, 1995). As Gormley remarks "(w)hereas education is widely regarded in positive terms, child care arouses much more ambivalent reactions. . . . The question of whether children under the age of six should be cared for outside the home continues to arouse fierce debate" (1995: 34). Cromley deals with the relationship between child care and education in her analysis of adoption of full-day kindergarten. In that context, she suggests that there is "tremendous downward pressure in the primary curriculum, including calls for public school instruction of 4-year-olds. At the same time, others are speaking to the perils of early academic demands" (p. 54). In short, US and Canadian policy makers currently face a dilemma that Deborah Phillips touches upon when she remarks that there is no agreement about whether child care is "a social intervention or an economic convenience; a service for children, for adults or for families; a comprehensive development program or basic caretaking; a supplement or a substitute for parental care" (1991: 165).

Ultimately, of course, child care plays a variety of roles. However, as Fincher points out in Chapter 10, the sense that there is a pressing need for government leadership in the field of child care in North America gets linked not only to the increased employment of mothers, but the increased employment of *middle-class* mothers. Politicians in both countries recognize that these women represent a significant segment of the voters and child care has become "mainstreamed" (Prentice, 1988) as a political issue and has become an important topic of government regulation and election promises (also see Phillips, 1991; Rose, 1990). Fincher argues that formal child care tends to privilege, and is geared toward white, middle-class families. Certainly, child care arrangements vary by the income level of parents: lower-income parents are more likely to rely on home-based care, especially care by relatives, whereas the children of better-off parents are more likely to be in child care centers (Burke *et al.*, 1994; Phillips, 1991). In part, this is due to formal care generally being more expensive than informal care, and because of limited subsidies. However, policy decisions about subsidizing child care mean that formal child care tends to cater to two

11

distinct groups: low-income, lone mothers and middle-class parents. Lone mothers often have lower rates of paid labor force participation and one reason is inadequate child care (Bloom and Steen, this volume; Léger and Rebick, 1993; McLanahan and Garfinkel, 1993). In Canada, about half of all formal child care spaces are subsidized and, as Truelove (Chapter 3) points out, the majority of the subsidies are used by low-income, lone-parent families. Moreover, most subsidized children attend non-profit centers (or municipal centers in Ontario, Québec, and Saskatchewan) (Pupo, 1991). Although this relationship is not as distinct in the US, the link has also been made between non-profit child care centers and attendance by children from low-income families (non-profit centers also have more ethnically diverse children) (Kamerman and Kahn, 1989).

Yet, as formal child care expands, it appears to be middle-class parents who are taking advantage of it (Melhuish and Moss, 1991b; Rose, 1990). For example, Holly Myers-Jones and Susan Brooker-Gross (Chapter 6) found that the majority of the parents in the centers they studied were in professional occupations. Damaris Rose (1993) reached a similar conclusion in her research on child care arrangements in Montréal. She also found that two-parent families with blue-collar jobs were much more likely to use informal care. Rose suggests that children who are white (or, in Canada's case, Canadian-born) are over-represented in child care centers (Rose, 1993; Rose and Chicoine, 1991). On the other hand, Rose also found that southern European immigrants often relied on extended families, while recent immigrants from the Caribbean (without extended family nearby) relied on non-relative, home-based, informal care. In the US, evidence also suggests that many African-American and Hispanic families depend on their extended family for care (Bean and Tienda, 1987; Farley and Allen, 1987; Klein, 1992).

Other aspects of child care in Canada and the US work to benefit some families more than others. Bloom and Steen, and Truelove (Chapter 3) point out that both countries have gradually increased tax deductions for child care costs. Indeed if tax credits are viewed as a type of subsidy, then they now constitute a larger share of federal spending on child care than subsidies through other avenues (including Title XX and AFDC in the US and CAP in Canada (but see note 2)). It is revealing that, as the debate around these deductions has evolved, the emphasis has been not on the notion of collective responsibility for children, but rather that child care expenses be acknowledged as a "business expense" for all working parents. Of course, this emphasis also reflects both the nervousness about "invading" family privacy and attempts to eliminate the "welfarist" stigma of publicly funded child care (Bloom and Steen, this volume; Friendly, 1994; Phillips, 1991).

Bloom and Steen raise an important point about tax deductions and the employment of in-home carers (like nannies) by middle-class families.

Following the 1989 decision that taxpayers' claims for the Child and Dependent Care Tax Credit must include the social security number of their child care provider, a number of President Clinton's Cabinet nominees admitted to hiring domestic help through the "underground economy." Indeed, it seems the nominees were not the only ones: Bloom and Steen suggest that as an apparent result of having to report social security numbers, claims dropped by more than 2 million. Moreover, while the majority of the informal child care sector in both countries consists of relatives and neighbors, there are the unknown, but significant numbers of illegal foreign workers.

Generally, however, tax deductions benefit higher-income families and are less beneficial to low-income families. For instance, in Canada there is no limit to the number of people who can claim child care tax deductions, but there is a limited supply of subsidized spaces, and, as Truelove (Chapter 3: 45) points out, "(t)he 1990s have seen a decrease in the number of subsidized child care spaces, and continued decreases in spaces and closures of centers are expected" (also see Mackenzie and Truelove, 1993). The shift toward tax deductions is represented as increasing parents' choice of child care. However, a number of commentators remark that it has really only increased choice for better-off parents and does not help those families who are unable to find child care in the first place (Cohen, 1993; Phillips, 1991; Mackenzie and Truelove, 1993; Rose, 1990). Increasing the amount of child care tax credits must be more broadly contextualized in terms of privatization and the rhetoric of "consumer preferences" (Kamerman and Kahn, 1989).

EVERYDAY GEOGRAPHIES OF WORKING MOTHERS

The earliest feminist geographers set about making visible the everyday geographies of women. They argued that women's activities and experiences of the urban environment were different from men's. Inequities and the constraints of gender roles for women were explored and documented, and it was argued that spatial structures helped to systematically disadvantage women relative to men. Particular emphasis was placed on sociospatial constraints, and time geography played an important role.[3] Feminist geographers argued that gender roles had to be acknowledged as a key social and spatial constraint for women, constraining their behavior, limiting their activities and confining them to a smaller geographic area than men. Jacqueline Tivers (1988: 84) argues that the "gender role constraint acts to restrict the activities of *all* women, whether or not they have children. It is women with young children, however, for whom the constraint is more physically obvious."

Over the last twenty years, several time geography studies have examined the spatial and temporal constraints of working women's daily activity patterns. Time geography studies highlight how women develop complicated schedules around the location and hours of work, the location and operating hours of child care facilities, transportation mode and availability, and the hours that their child(ren) need(s) child care (Bowlby, 1990; Droogleever Fortuijn and Karsten, 1989; Palm and Pred, 1978; Pickup, 1984, 1988; Tivers, 1985, 1988; Michelson, 1985, 1988). The continued usefulness of time geography to describe mothers' everyday geographies is reflected by Dyck, England, Myers-Jones and Brooker-Gross, and Truelove (Chapter 7), who all discuss time geography to varying degrees (however, see G. Rose, 1993: Ch. 2, for a feminist critique of time geography).

The sociospatial constraints theme also percolates through the literature exploring gender differences in commuting. Regardless of the place or time period examined, women generally have shorter commutes than men, both in terms of time and distance. That women are traditionally responsible for the home and children emerged as one of the earliest explanations as to why women have shorter work-trips than men (see Blumen, 1994, for an overview of the literature covering various explanations for gender differences in commuting). While the ages and number of children affect the probability of a woman being in the paid labor force (as well as whether she works full- or part-time), the same does not apply to men; and working women, unlike working men, are more likely to trade off commuting time to accommodate their household responsibilities (Bowlby, 1990; Hanson and Pratt, 1988, 1990; Johnston-Anumonwo, 1992; Pickup, 1984, 1988; Pratt and Hanson, 1991a; Singell and Lillydahl, 1986). On the other hand, there is mixed empirical evidence regarding the "common sense" explanation that working mothers have shorter commutes than childless women. Some studies conclude that the presence of young children reduces women's commutes (Ericksen, 1977; Fagnani, 1983; Pickup, 1984; Preston *et al.*, 1993; Singell and Lillydahl, 1986); while others find that mothers have longer or similar commutes to childless women (Gordon *et al.*, 1989; Hanson and Johnston, 1985; Johnston-Anumonwo, 1992).

There have been very few studies that look specifically at journeys to *child care*. The overwhelming majority of journeys to child care are undertaken by women (Myers-Jones and Brooker-Gross, this volume; Michelson, 1985, 1988; Rosenbloom, 1988, 1993). For example, Rosenbloom (1993) found that married women make more linked work-trips (for instance, their commutes include journeys to child care) than either married men or lone mothers (who are more reliant on public transportation than married mothers). Michelson (1985) found that 70 per cent of journeys to child care involved the mother, either alone (55.6 per cent) or accompanied by the father (14.5 per cent). He noted that the journey to child care is usually shorter than the parent's trip from child care to workplace, but journeys to child care lengthen commutes by approximately 28 per cent, and that this "extra" travel is particularly pronounced in the work-trips of women with preschool children. Myers-Jones and Brooker-Gross and Truelove (Chapter 7) add to our knowledge of journeys to child care. Both studies conclude that journeys to child care are short. Myers-Jones and Brooker-Gross examine the journey to child care in a non-metropolitan setting, dominated by the Virginia Polytechnical and State University. They do not simply consider distance, but place the journey to child care within the broader framework of families' coping strategies and the interrelationships between the geographies of child care, housing and employment. Truelove (Chapter 7) focuses on the "mixed economy" system of child care in Canada in her case study of journeys to child care to different types of formal child care centers in Metropolitan Toronto (non-profit, commercial and municipal). Most subsidies are given to low-income families and/or lone mothers, and Truelove finds that regardless of the type of center attended, subsidized children have shorter trips than full-fee children. She also finds that full-fee suburban children travel farther to centers than similar children in the city, possibly reflecting the more dense network of child care centers in the city. However, Truelove does not find a suburban–central city difference for children with subsidies, and suggests that this reflects the government's tendency to encourage families with subsidies to attend their nearest center.

Another set of literature focuses on the everyday geographies of working mothers against the backdrop of the socially constructed nature of gender relations in the context of the gender division of labor and the separation between spaces dedicated to production and reproduction. Rather than emphasize constraints, women's experiences are located within the context of the mutually reinforcing relations between production and reproduction, and capitalism and patriarchy. The interconnections and linkages between the public and private, home and (paid) workplace are emphasized, as is the fluidity of the everyday as women (and men) weave together their daily activities and responsibilities (Bowlby, 1990; Hanson and Pratt, 1988, 1990; Mackenzie and Rose, 1983; Pratt and Hanson, 1991a). For instance, Suzanne Mackenzie (1987) placed the increased demand for home-based child care (provided by one group of women) in the context of the restructuring of the British Columbia economy, as another group of women, who had previously been full-time homemakers, took on paid jobs to supplement their family's income. Dyck, England, and Myers-Jones and Brooker-Gross all focus on juggling various responsibilities within the web of home, child care and workplace, while acknowledging the importance of broader socioeconomic relations.

Recently, some feminist geographers have moved away from casting women as passive receivers of wider social and economic relations, stressing the transformative and creative capacities of women as knowledgeable, strategizing agents, possibly able to change their sociospatial systems in order to better fit their needs (Dyck, 1989, 1990, this volume;

England, 1993a, this volume; Pratt and Hanson, 1993). This is not to deny that women face structural and spatial constraints. Clearly, such constraints are significant and may operate without people's awareness of them, but they do depend on human agency to be reproduced. England examines the strategies that suburban working mothers in Columbus, Ohio, employ to cope with a multiplicity of overlapping and often contradictory roles and spatial factors. She emphasizes their ability to create and adapt within a pre-existing and evolving web of localized relations. Dyck examines the strategies that suburban mothers in Vancouver, British Columbia, create to manage the practical and moral dilemmas of combining paid employment with motherhood. She finds that as active shapers of their worlds, the mothers' strategies are embedded in their concept of "safe space" and "good mothering," which, in turn, are defined and negotiated in the context of their neighborhood as well as broader social and economic changes.

THE STRUCTURE OF *WHO WILL MIND THE BABY?*

The body of the book is divided into four parts. Each part provides examples from both the US and Canadian contexts. The first two parts deal with the government's role in child care policies and child care provision in both countries, with consideration of the national, state/provincial and local levels. In Part IV, the focus shifts to the local scale, with attention centered on the everyday geographies of working mothers, in terms of the journey to child care and working mothers' coping strategies. Part V of the book places child care in international perspective.

David Bloom and Todd Steen (Chapter 2) and Marie Truelove (Chapter 3) examine government policies, funding and provision of child care in the US (Bloom and Steen) and Canada (Truelove). Bloom and Steen focus on the labor force implications of increased social investment in the "child care industry." They explore how the expansion of the child care industry could remove a major barrier to paid employment for a sizeable number of women, especially women in low-income and lone-parent families. Bloom and Steen argue that expanding the child care industry will also help employers reduce labor turnover and increase productivity. In her first chapter in the book, Truelove examines the emergence of child care as a prominent issue on the Canadian political agenda. In Canada limited funding for child care is shared by the federal and the provinces/territories (and in the case of Ontario, local government), while delivery and regulation are the responsibility of the provinces and territories. Truelove explores how these policies are played out in the specific context of Metropolitan Toronto.

The impact of intra-state/provincial variations in social and economic characteristics on child care provision is further explored in Part III of the book. Ellen Cromley (Chapter 4) provides a US case study and Ian Skelton

(Chapter 5) contributes a Canadian example. Ellen Cromley investigates the connections between community characteristics, early childhood education and the availability of full-day kindergarten in the state of Connecticut. She distinguishes between those communities where full-day kindergarten is available and those where it is not. Cromley contends that something other than the labor force participation of mothers has motivated communities to adopt a full-day program. Interestingly, in light of current debates regarding the educational component of child care, she argues that a community's commitment to preschool education (along with high median family income) is most strongly related to full-day kindergarten availability. Turning to the question of spatial equity in child care provision, Ian Skelton traces the roots of spatial differences in child care provision within Ontario. He argues that public policies governing the funding and delivery of child care services do not support the development of intra-provincial equity in provision and, in fact, militate against it. Skelton then assesses the patterns of provision over the post-Second World War period and analyses the contemporary intra-provincial differences in supply. Given concerns over the extent to which child care provision privileges middle-class families, Skelton finds that although service levels *are* associated with measures of need, they are more strongly responsive to socioeconomic status.

Moving to the local scale, the chapters in Part IV of the book focus on how the different locations of home, workplace and child care facilities frustrate the everyday geographies of working parents. Holly Myers-Jones and Susan Brooker-Gross (Chapter 6) and Marie Truelove (Chapter 7) examine the journey to child care. Myers-Jones and Brooker-Gross place the journey to child care in the context of parents' domestic arrangements and coping strategies in Blacksburg, Virginia, while Truelove explores it in light of government policies in Metropolitan Toronto. As there is limited data available to conduct research on this topic, the authors generated their own child care center-based data: Myers-Jones and Brooker-Gross surveyed parents, while Truelove obtained address lists. Myers-Jones and Brooker-Gross look at formal child care in Blacksburg, Virginia – a small city with few choices in either child care or paid employment. They find that child care trips are primarily the mother's responsibility, and that they tend to be short in distance and related to both distance to home and to work. As Blacksburg is a small city and has a shortage of available child care, Myers-Jones and Brooker-Gross expected to find limited sensitivity to distance. However, one third of their respondents indicated that distance from home was an important factor in their decision-making process. The primary reasons that parents chose the center they used were the quality of care and the educational value of the center. Myers-Jones and Brooker-Gross explore the socioeconomic status of the parents and find that over half of the respondents are in professional occupations. This figure is high partly because the university is the largest employer, but it is also because relatively well-paid professionals are better able to afford the cost of formal child care. Myers-Jones and Brooker-Gross' results reflect the issues of the educational benefits of child care and the increased association between middle-class parents and formal child care raised by Cromley and Skelton.

Marie Truelove's second chapter in the book explores the changing geography of Metropolitan Toronto's child care centers and the travel patterns of children that attend them. She finds that different municipalities are dominated by particular forms of child care provision. In terms of the journey to child care, she differentiates between children attending three types of child care centers (commercial, municipal and non-profit), pays particular attention to the trips of full-fee versus subsidized children, and to differences among the municipalities. While Truelove finds that the mean distances from home to child care are short, they are especially short for subsidized children whose families are encouraged to use centers close to their homes. She notes that the distances traveled vary by type of child care center and the municipality within Metropolitan Toronto. Moreover, she finds that not only do most children not attend their nearest center, but that this is not explained by cost differentials. Instead, Truelove suggests, it is connected to factors such as the availability of spaces, parental preference and the proximity to the workplace.

Women who combine their roles as mothers and paid workers face complicated schedules: coordinating trips to child care and workplace, in addition to maintaining their homes and families. Two chapters are based on case studies of suburbs. Traditionally, suburbs have been associated with family-centered households and full-time mothers. Increasingly, suburban women are in paid employment, and Kim England (Chapter 8) and Isabel Dyck (Chapter 9) explore the strategies that suburban women employ to cope with their multiple roles. Both used qualitative methods in order to reveal women's organization of, and strategies for coping with combining motherhood and paid work. England interviewed working mothers in middle-class suburbs of Columbus, Ohio, who are all white, but differ by age and marital status. She also interviewed employers to examine their responses to dilemmas generated by employing women with young children. England concentrates on a broad definition of coping strategies and examines women's attempts to alter their sociospatial systems in order to better negotiate their multiple roles. In particular she explores how women deal with the stresses associated with the daily transition between home and work, their coping strategies for dealing with combining motherhood and paid work, as well as those associated with domestic chores and child rearing. England draws on interviews with employers to consider their attempts to create "family-friendly workplaces" in terms of child care benefits and flexible work schedules.

Isabel Dyck draws on accounts of white, married women of varying socioeconomic statuses living in a Vancouver suburb. Dyck specifically focuses on women's use of informal child care to solve the problem of limited formal child care. Dyck argues that faced with a dilemma in contradictory values concerning mothering and wage labor, women create informal child care solutions that are culturally sanctioned in the context of the local setting. She argues that a woman's own interpretation of what constitutes "good" mothering impacts upon the precise form of child care solutions that she develops. Dyck further emphasizes the centrality of localized sets of relations for providing information, support and solutions to child care problems.

In Part V, child care is placed in international perspective by Ruth Fincher (Chapter 10). She provides a springboard into areas for future research with regard to state interventions in child care provision. Fincher outlines four major themes that geographers should consider for further research. The first relates to the state's philosophies of, and assumptions about women's roles in child care provision. The second focuses on representations of the child care user in government provisions, the implication being that those people who fit that image are better served than those who do not. The third theme deals with the role that child care activists have played in shaping public policy. The final theme for future research looks at the diversity of forms of government child care provision. These themes are explored using examples from a wide range of countries, with particular attention paid to the distributional implications for women of diverse characteristics.

Part II

CHILD CARE POLICY IN THE UNITED STATES AND CANADA

2

MINDING THE BABY IN THE UNITED STATES

David E. Bloom and Todd P. Steen

INTRODUCTION

During the past thirty years the labor force activity of women in the US has undergone remarkable change. The 59 million women in the labor force in July 1993 – 45 per cent of the total civilian labor force – represented more than double the number of women in paid employment in 1965. However, even this large increase in the number of women in the labor force pales in comparison to the increased labor force activity of women with preschool-age children. Almost 59 per cent of all mothers with children under the age of 6 are currently in the labor force (that is, employed or unemployed but looking for work), compared to less than 22 per cent in 1960 (Table 2.1). In 1993, 59.6 per cent of married women with children under the age of 6 were participating in the labor force, compared to 18.6 per cent in 1960.

Despite the rapid entry into the US labor market of women with young children, the scarcity and high cost of quality child care prevent many mothers from working. Numerous other industrial nations provide a far more generous set of child care and maternity leave benefits than does the

Table 2.1 Labor force participation rates of women with children under age 6, by marital status, for selected years (%)

Year	Overall	Married	Single	Other
1960	21.6	18.6	N/A	40.5
1970	33.2	30.3	N/A	52.2
1980	47.4	45.1	44.1	60.3
1985	53.7	53.4	46.5	59.7
1990	58.5	58.9	48.8	63.6
1993	58.3	59.6	47.4	60.0

Notes: Women are 16 years old and above. The category 'other' includes widowed, separated, and divorced women. N/A – not available

Source: US Bureau of the Census (1994a)

US. The expansion of the US labor force during the last thirty years, caused by the combined effects of the baby boom and the influx of women into the labor market, allowed public policy makers and employers to pay relatively little attention to the many women who wanted either to work or to work more hours but could not because of family responsibilities. These women, however, constitute a large segment of the adult population and their employment needs will demand increased social attention, especially as labor markets tighten through the end of the century.

The size of the US labor force is no longer increasing as fast as it was throughout most of the period since the Second World War, and the workers' average age is increasing as the baby boom generation grows older. At the same time, employers are searching for workers qualified to perform increasingly complex types of work. These difficulties may become even more serious in the future as the "baby bust" generation (children born between 1965 and 1976) continues to enter the labor force. By 2005, the number of workers aged 20–24 will be 3 per cent less than in 1979, while the number of teenaged workers will be 8 per cent less (Fullerton, 1993).

Many possible policy responses, both public and private, could improve employers' ability to recruit, retain and motivate workers. These include flexible scheduling, increased use of computers and other machinery in place of workers, and relaxed immigration requirements. However, none of these responses is better suited, at least in principle, to satisfying the needs of both employers and households, while simultaneously promoting a variety of national interests, as policies that are aimed at expanding the child care industry and improving the quality of care. Such policies would enable many parents, especially women, to enter the labor force, to work more hours, and to enter more demanding and career-oriented occupations (see Bloom and Steen, 1990). In addition, improved child care policies could lead to better morale, higher productivity and reduced turnover among workers, thereby enhancing the competitive position of US businesses. Expanding the child care industry might also ease the strain on government budgets caused by welfare payments made to many mothers not in the labor force. By making it easier for women simultaneously to work and care for their families, expanded child care might also help to boost flagging fertility rates, and thereby mitigate problems associated with the projected "birth dearth" (Bloom, 1986; Wattenberg, 1987).

Expansion of the child care industry will not be achieved easily, however. Child care workers are often young and are typically paid low wages, meaning that attracting new workers into this industry during a period of slow labor force growth is especially difficult. Although some employers are now providing employees with child care assistance, with benefits ranging from on-site child care centers to child care resource and referral services, employer initiative in the area of child care is still in a state of infancy. The expansion of government funding for child care also seems unlikely in a time of large budget deficits at both the federal and state levels and widespread anti-tax sentiment. Nonetheless, as employers and public policy makers begin to recognize the potential benefits of expanded child care, the growth and development of this industry can be expected to accelerate.

US GOVERNMENT POLICIES TOWARD CHILD CARE

Public policy makers in the US have taken some actions during the past few decades to ease the private burden of child care costs, but these have been relatively limited compared to the policies of many other industrial nations (see Fincher, this volume). Until 1990, several attempts to pass a comprehensive package of child care legislation in the US had failed, leaving a patchwork of direct and indirect programs at the federal, state and local levels (for an example, see Cromley, this volume).

The US federal government's first involvement with child care was in 1933. During the Great Depression, funds were provided under the auspices of the Works Progress Administration to establish child care centers and nursery schools. Although the program's primary purpose was to provide unemployed teachers with jobs, this plan marked the first time federal monies were used to support child care. By 1937, around 40,000 children were enrolled in child care provided through the Works Progress Administration (Reeves, 1992). The Second World War provoked even greater government involvement in child care provision. The Lanham Act of 1941 disbursed more than $51 million in funds for child care centers to be used by mothers working in defense industries. This period saw the establishment of more than 3,000 centers serving 600,000 children. In 1946, however, with the end of the war, all federal funding for child care was terminated.

In 1970, the White House Conference on Children identified child care as a major difficulty for families in the US. The growing interest in child care led Congress to pass the Comprehensive Child Development Act in 1971. This legislation designated $2 billion for child care funds, including monies for helping working-class families with child care needs, providing child care for welfare recipients, and funding for the development of new child care resources. However, President Nixon vetoed the legislation, labeling it as "family weakening," thereby effectively halting efforts to extend Federal aid for child care until the late 1980s (Melhuish and Moss, 1991a).

Significant financial help to parents came in 1954, when an amendment to the tax law allowed lone-parent families and low-income, two-earner families to take a tax deduction for certain types of child care expenses. In 1976, the federal government replaced this deduction with the Child and Dependent Care Tax Credit. As it currently exists, it allows a family to decrease its annual tax liability by up to $720 for one child or $1,440 for two or more children. Families with annual incomes below $10,000 can claim a tax credit of 30 per cent of child care expenses up to $2,400 ($4,800 for two or more children).[1] This percentage drops steadily from 30 per cent to 20 per cent for families with incomes of $30,000 or more. However, the credit cannot be larger than the family's tax liability, which effectively reduces the benefit for many low-income families, and so primarily benefits middle-income families. In 1990 just 18 per cent of families using the credit had annual incomes of less than $20,000. The Tax Reform Act of 1986 preserved the Child and Dependent Care Tax Credit, making it one of the few large deductions still available to federal taxpayers. However, beginning in 1989, taxpayers who claimed this credit were required to report the social security numbers of their child care providers. As an apparent result, the number claiming this credit dropped from 8.7 million in 1988 to 6.1 million. Estimates from the Internal Revenue Service indicate that an extra $1.2 billion of taxes were collected due to the decrease in the number of taxpayers who claimed the credit. Clearly, child care providers are a significant part of the underground economy in the US. This received a great deal of media attention in 1992, when several of President Clinton's Cabinet nominees admitted that they had not paid social security taxes for child care workers and other household help.

In their assessment of the impact on child care of President Reagan's administration (that encouraged privatization), Kahn and Kamerman (1987b; Kamerman and Kahn, 1989) estimated that between 1980 and 1986, *direct* federal funding for all child care programs (as opposed to tax credits) decreased by 18 per cent in inflation-adjusted dollars, from $2.5 billion to $2.1 billion (in 1986 dollars). Funds allocated to states under Title XX of the Social Security Act of 1975 are a major source of direct federal subsidies for child care. Nominal expenditures on child care under this program fell from $600 million in 1980 to around $400 million in 1986. These declines were especially severe given the concomitant 29 per cent increase in the number of working mothers with preschool-age children. Between 1981 and 1985, many state governments also decreased the nominal amount of funds independently allocated for child care programs; in inflation-adjusted terms, nearly all states reduced child care funding. In addition, many states lowered their standards for the quality of care that centers must meet to be eligible for support under this program, while simultaneously relaxing their quality enforcement policies.

In 1988, in response to growing concerns about the declining quality and availability of child care in the US during the 1980s, more than 100 pieces of legislation dealing with child care were introduced into the US Congress. Senators Christopher Dodd (Democrat, Connecticut) and Orrin Hatch (Republican, Utah) launched a bi-partisan attempt to pass a comprehensive child care legislative package, the Act for Better Child Care Services (the ABC bill of 1988). This bill proposed both an increase in direct federal grants to states for child care (along with a requirement for states to develop better standards for child care) and an expansion of tax credits for working parents. Substantial differences in the House bill and the threat of a presidential veto resulted in this legislation never passing both houses of Congress and reaching President Bush's desk.

Continuing interest in the issue of child care led to Congress passing a substantial child care package in the 1990 Omnibus Budget Reconciliation Act. The primary objective of this legislation was to help low-income Americans. It included four major elements: (1) the expansion of the Earned Income Tax Credit, (2) block grants to states to help improve their availability of child care, (3) grants to those working poor not already on Aid to Families with Dependent Children (AFDC) and (4) funds for states to upgrade their licensing requirements for child care facilities. These monies have made center-based care more affordable (Gormley, 1995; Reeves, 1992).

Although not directly tied to the use of child care, the Omnibus Budget Reconciliation Act provided $12.5 billion over five years to expand the Earned Income Tax Credit. Initially established in 1975, this credit primarily benefits low-income working parents who can receive a tax credit which declines as earnings increase (the Earned Income Tax Credit was expanded again in 1993). In 1994, the credit for eligible families with one child was $2,038 for taxpayers earning less than $11,000. Additional amounts are available for those families with more than one child or a child under 1 year old. The money does necessarily go to pay for non-home-based child care, for example, an estimated $5.2 billion of the total funding for this program is used to finance a credit for parents purchasing health insurance for their children. Unlike the Child and Dependent Care Tax Credit, one advantage of the Earned Income Tax Credit is that it is available to taxpayers regardless of their total liability (so if the claim exceeds income tax liability, the balance is reimbursed).

Under the second part of the Act – the Child Care and Development Block Grants – the government allocated $2.5 billion for three years to states in the form of block grants to be used to improve both the quality and the availability of child care (funding for Block Grants was extended in 1993). Of this money, 75 per cent is allocated for families with incomes below 75 per cent of the median state income. This money will be directed to either the families themselves, or to the providers in the form of vouchers. The remaining 25 per cent is allocated to latch-key programs, monitoring activities and other programs. The third element of the Act – the At Risk program – provides an additional grant of $1.5 billion for child care subsidies under Title IV-A of the 1975 Social Security Act. These funds are earmarked for low-income families who do not qualify for welfare, and are intended to help families "at risk" from backsliding into AFDC dependency. The fourth part of the Act provides a small amount in grants for states to improve licensing and registration requirements for child care centers.

A further piece of legislation, the Family and Medical Leave Act, passed in February 1993, also achieved some progress in the care of the youngest children in the US. Under this legislation all employers with fifty or more employees are required to provide up to twelve weeks a year of unpaid leave to an employee upon the birth of a child. The employer is also required to guarantee that the employee will be able to return to work at a similar job, and must continue the employee's health benefits during the leave. To be eligible for this benefit the employee must have worked for the employer for at least one year, and have worked an average of twenty-five hours a week during the previous twelve months. Any state law that provides a more generous leave policy remains in effect and is not displaced by the federal legislation.

This additional child care legislation and parental leave policy still leaves the US significantly behind compared to the policies of Canada and Europe (see Truelove, Chapter 3, and Fincher, this volume). The provision of child care in the US is left primarily up to the private sector, and its regulation is often lax. Even though parental leave can greatly facilitate child care for infants, parents in the US are still forced to give up their paychecks if they want to take advantage of this option.

CURRENT TRENDS IN CHILD CARE USE

The types of child care that Americans use have changed noticeably since the 1970s. Given the lack of government involvement in the provision of child care in the US, the private sector provides the vast majority of child care. Nationally representative surveys are only collected intermittently. A comparison between two such surveys, the June 1977 Current Population Survey and the 1991 Survey of Income and Program Participation, highlight a trend toward the use of more formal child care facilities such as nursery schools and child care centers (US Bureau of the Census, 1994b). In 1991, about 23 per cent of mothers with children under age 5 used a child care center or preschool as their primary arrangement, compared with 13 per cent in 1977. Another 36 per cent of mothers used a relative, neighbor or friend to care for their children at home, up from 34 per cent in 1977. In 1991 only 9 per cent of mothers cared for their youngest child while they worked, compared with 11 per cent in 1977. (Recently released, but incomplete figures for fall 1993 indicate a continued trend toward working mothers with children under age 5 using a child care center or preschool (29.9 per cent) and fewer (6.2 per cent) caring for their youngest child while they worked (Casper, 1995a).)

The 1994 US Bureau of the Census study also showed that types of child care working mothers use varies according to their marital status (see Table 2.2). In 1991, husbands provided child care in 22.9 per cent of households where both parents are present and where the wife worked for pay. By contrast, only 7 per cent of the mothers in the "other marital statuses" group (e.g. divorced) relied on the child's father. Lone mothers compensated for the limited availability of the father by greater use of grandparents (24.8 per cent versus 13.7 per cent for married mothers with spouse present) and other relatives (11.6 per cent versus 6.7 per cent for married mothers with spouse present) (US Bureau of the Census, 1994b).

Table 2.2 Distribution of primary child care arrangements used by working mothers, for children under age 5, by marital status, June 1977 and Fall 1991

Married, husband present (%)	1977	1991
Total	100.0	100.0
Care in child's home	34.4	36.6
by father	17.1	22.9
by grandparent	[**]	5.6
by other relative	10.1	2.5
by non-relative	7.2	5.5
Care in another home	40.1	29.5
by grandparent	[**]	8.1
by other relative	17.6	4.2
by non-relative	22.6	17.1
Organized child care facilities	11.6	22.7
Mother cares for child	12.9	9.8
Other arrangements	0.9	1.4

All other statuses (%)	1977	1991
Total	100.0	100.0
Care in child's home	31.3	31.7
by father	0.8	7.0
by grandparent	[**]	14.1
by other relative	24.8	6.0
by non-relative	5.7	4.5
Care in another home	43.4	37.6
by grandparent	[**]	10.7
by other relative	21.6	5.6
by non-relative	21.8	21.3
Organized child care facilities	19.1	24.5
Mother cares for child	4.4	3.7
Other arrangements	1.6	2.5

Notes: Women are 18 to 44 years old.
The category 'All other statuses' includes widowed, divorced, separated and never-married women.
[**] In 1977, percentages for "other relative" category include care by grandparents

Source: US Bureau of the Census (1994b)

Child care arrangements also vary by women's employment status. Mothers working full-time tend to use organized child care facilities almost twice as often as mothers working part-time (28 per cent versus 15 per cent in 1991). Children of part-time workers are more likely to be cared for in their homes than children of full-time workers (46 per cent versus 30 per cent) or to be cared for by the mother while she works (15 per cent versus 5 per cent) (US Bureau of the Census, 1994b).

It is difficult to obtain comprehensive, up-to-date figures for the cost of child care. However, evidence indicates that the prices vary widely depending on the type of child care and location. Among the most recent estimates are those prepared by the Children's Defense Fund, which indicate that the 1991 average yearly cost of full-time child care was $3,432 per child. And Census data for 1991 indicate that the average cost per hour for child care provided by a relative was $1.88, while the cost per hour for center care averaged $2.15 (US Bureau of the Census, 1994b). The most expensive care child care arrangements are those where the child is cared for by a sitter in the child's own home. Some households, however, have the option of having a grandparent or other relative provide child care at little or no pecuniary cost. The cost of child care varies across the country: in 1991, it averaged $8,944 per year in New York City and $5,668 in Boston, compared to only $2,964 in Miami (Gormley, 1995; Reeves, 1992). In 1993, child care was more expensive in metropolitan areas ($80 per week), compared with non-metropolitan areas ($55 per week); and families in the northeast paid more for child care ($85 per week) than those in the midwest and south (about $70 per week) (Casper, 1995b). The cost of child care makes it the fourth largest household expense after housing, food and taxes. Moreover, recent evidence suggests that child care is more of an economic burden for low-income families, in that it represents a larger portion of their budget (18 per cent in 1993), than for non-poor families (7 per cent) (Casper, 1995a).

THE CHILD CARE INDUSTRY

As women's paid labor force participation rates have grown, so has the need for all forms of child care. The child care industry is a small but growing segment of the service sector of the US economy. The industry comprises a varied and rapidly changing set of activities and organizations, ranging from a mother caring for her own and her neighbor's children at home, to large corporations operating hundreds of child care facilities across the country. Child care resource and referral services, a relatively recent development, help parents locate and choose appropriate caregivers for their children. This section of the chapter provides background on both supply and demand factors related to the child care industry in the US.

More than 9 million preschoolers spend at least part of their day being cared for by someone other than their mothers. According to recent estimates prepared by the National Research Council (see Hayes *et al.*, 1990), the US had more than 64,000 licensed child care centers in 1988. In comparison, a national survey conducted in 1977 by Abt Associates reported only 20,000 such centers. In addition, in 1988, some 200,000 family homes were registered to provide child care, up from 115,000 in 1977 (see Kahn and Kamerman 1987b). These licensed homes are almost certainly a minority of all the households that receive some payment for providing child care (Hayes *et al.*, 1990; Kahn and Kamerman, 1987b).

The Bureau of Labor Statistics estimates that there were 1.23 million child care workers in the US in 1987, up 65 per cent from the 727,000 child care workers estimated in the 1980 Census of Population, and up more than 200 per cent from the 395,000 workers in 1970. Of these workers, 33 per cent (405,000) are employed as private household workers, mainly as babysitters or caregivers in the child's home. By contrast, in 1970, more than 55 per cent of child care workers were employed in private households (US Bureau of Labor Statistics, 1990). Clearly, this reflects the growth in center-based child care. These trends also show the tremendous growth in the numbers of child care workers. Recent studies rank child care workers as the thirteenth fastest growing occupation for 1992–2005 (US Bureau of the Census, 1994a).

Child care is a highly sex-segregated occupation: more than 96 per cent of child care workers are women. As concerns race, 12 per cent of child care workers are African-American and 8 per cent are Hispanic (the population as a whole is 10 per cent African-American and 7 per cent Hispanic) (US Bureau of Labor Statistics, 1990). At the same time, wages in the child care industry are quite low. According to data from the 1989 National Child Care Staffing Study (cited in Reeves, 1992), the average full-time child care worker, with a high school education or less, made $8,120 per year, compared with an average of $15,806 for the typical woman with the same level of education. Child care workers with a college degree fared even worse: full-time wages were $11,603, compared to an average of $26,006 for all college-educated women. Relatively low wages presumably explain why the annual labor turnover rate in the child care industry is 48 per cent, well over double the national average of 18 per cent.[2]

Clearly, expanding the supply of child care services requires the recruitment of additional workers to the labor-intensive child care sector. To attract more child care workers, wages will have to rise and working conditions will have to improve, especially in tight labor markets. As the wages of child care workers typically amount to between 60 and 80 per cent of the total costs of running a child care center (more for family day care homes), higher wages would raise the cost of child care, making it even more difficult for many to afford, especially lower-income families. In other words, for women with low incomes, financial assistance from both government and employers may be essential if they are to have the increased access to affordable child care that will permit them to increase their labor market activity.

EMPLOYER POLICIES TOWARD CHILD CARE

Relatively few US firms currently provide child care services or benefits directly to their employees. Table 2.3 reports that in 1991 only 8 per cent of all full-time employees in medium and large establishments (those employing 100 or more workers) were eligible to receive child care benefits.[3] This does, however, represent an increase from the 1 per cent eligible for such benefits in 1985. In 1990, 9 per cent of state and local government employees were eligible for employer assistance for child care, whereas, only 1 per cent of employees of smaller, private firms were eligible for such benefits. However, other surveys do exist that suggest that employer assistance for child care is offered to a somewhat greater degree than these data indicate. This disparity arises because of differences in the type of employee benefit considered to be assistance for child care (see for example, Hofferth *et al.*, 1991).

Employers have a wide range of options for providing child care assistance, all of which entail advantages and disadvantages for both businesses and workers. About 1,800 employers currently provide on-site or near-site child care services (Reeves, 1992).[4] These centers have the advantage of allowing employees to be close to their children, enabling parental visits during lunch time or coffee breaks. A disadvantage of on-site care is that in many cases young children may have to accompany their parents during long and difficult commutes to and from work.

The advantages to companies of providing on-site child care include increased productivity as it may alleviate many employees of concerns that can inhibit their work performance. On-site child care may also contribute to reduced worker turnover by effectively increasing both the pecuniary and

Table 2.3 Full-time employees eligible for specified benefits, by employer type, 1990 and 1991 (%)

Benefit	Medium and large firms, 1991	Small firms, 1991	State and local governments, 1990
Child care	8	1	9
Elder care	9	2	4
Adoption assistance	5	1	1
Long-term care insurance	4	1	2

Notes: Medium and large firms have at least 100 employees; small firms have less than 100 employees

Source: US Bureau of Labor Statistics (1991, 1992, 1993)

non-pecuniary costs for employees if they change jobs. However, providing on-site child care can be expensive and demand for this service fluctuates over time. Liability issues may also be involved. In addition, this benefit may be perceived as discriminatory, because it is typically only useful to a small fraction of a firm's employees (Kahn and Kamerman, 1987b).

An increasingly important method through which employers can provide child care assistance to employees is flexible spending accounts financed by salary reduction (see Bloom and Trahan, 1986). Under this provision of federal tax law (Employer-Based Dependent Care Assistance Plans), employees can reduce their salaries by an amount specified at the beginning of the calendar year and use that money to fund a child care reimbursement account that is used to cover eligible child care expenses incurred during the year. Thus, child care expenses are effectively paid from pre-tax earnings, a significant discount for many employees, but especially those in the highest income brackets. This mechanism is relatively easy for employers to administer, and can even generate savings for the employer in the form of reduced social security and unemployment taxes. Under the present tax law, employees must forfeit to their employer any money not spent on child care by the end of the calendar year, so employees do assume some risk when they establish a flexible spending account. The maximum salary reduction allowed for child care is $5,000 per year. This amount has not been increased since it was established in 1981 and it is currently well below the annual cost of child care in several large metropolitan areas of the US (as noted earlier).

During the 1980s, there was rapid growth in the use of resource and referral agencies by employers to assist their employees with child care. Resource and referral agencies provide a variety of services, including increasing the local supply of child care by training local providers. Currently, around 500 agencies exist across the US, up from some sixty in the early 1980s (Gormley, 1995; Reeves, 1992). When firms use these services, information on the quality, cost and location of child care options are typically provided to employees at no charge, but employees pay for the child care themselves. In some cases the agencies also conduct on-site seminars to help employees deal with various child care issues. These resource and referral services may be offered on an in-house basis or under contract with an outside firm or agency. This employee benefit is usually relatively easy and inexpensive for the firm to supply.

Companies can also assist their employees with child care in a number of other ways. Many employers allow employees to work at home, giving them greater flexibility if child care arrangements break down or if child care is needed for short periods of time. Some firms reserve spaces in local child care centers and make them available to their employees, or provide a voucher for all or part of the cost of care at certain child care centers. Other options include flexible working schedules, job sharing or programs for sick children.

Child care assistance is still a relatively rare fringe benefit, but its prevalence is increasing. Assistance with child care has the potential to increase both workers' productivity and their loyalty to the firm. However, when calculating the gains that accrue from such policies, firms may underestimate the benefits to the firm itself, and especially those to the worker and to society as a whole. Government action to encourage firms to provide child care benefits could result in substantial long-term benefits for all parties involved.

CONCLUSIONS

This chapter examined some of the paid labor force implications of expanding the child care industry in the US. The findings support the view that additional resources should be devoted to this industry. Other studies have estimated the actual dollar returns on additional investments in child care. According to the Children's Defense Fund, every dollar invested in high quality child care yields $6 in reduced spending on other government programs. The House Select Committee on Children, Youth, and Families reported that each dollar spent on preschool education yields a return of $4.75 in reduced costs for special education and welfare. So investment in such programs makes good economic sense (Reeves, 1992).

The findings of this chapter provide some limited guidance as to the appropriate source of additional investment in the child care industry: should it be the government, employers or families? For example, many of the supposed benefits to employers are also benefits to employees, including less absenteeism and less stress associated with worrying about the safety and well-being of children while at work. This suggests that the burden of increased investment in the child care industry should not be placed entirely upon the shoulders of employers. However, the reduction in the growth rate of the US labor force may catch many employers off-guard and prevent them from fully reaping the potential benefits of expanded child care. Thus, while market forces may eventually lead US businesses to invest more in child care and to enjoy the resulting benefits, the government might perform a useful short-run function by demonstrating these benefits through its own personnel policies and by encouraging and helping employers to plan for the future.

The lack of information on the long-run effects of child care on the development of children's social and cognitive skills might also provide a positive basis for government intervention in the child care industry. The current generation of US children is the first to be cared for so extensively by individuals who are neither their parents nor other relatives. A great deal of supervision and care previously provided by mothers is now provided by a cadre of child care workers whose talents, training and commitment to children vary widely. As the government has a legitimate interest in safeguarding and promoting the well-being of its future adults, increased investment in the child care industry, especially in the quality of child care, may be justified on the grounds that it will reduce the risks inherent in present arrangements, and may also yield a large future return to society. Much additional research is needed, however, before these issues can be resolved in anything approaching a definitive manner.

3

MINDING THE BABY IN CANADA

Marie Truelove

INTRODUCTION

The growth of interest in, and demand for the provision of child care in an era of declining fertility rates might seem to be paradoxical, but it is not (see Becker, 1981, for an economic analysis of fertility rates). The demand is generated primarily by the increase in the paid employment of mothers of young children; mothers with one or two children are more likely to work outside the home than mothers with larger families. The increase in the paid labor force participation of women – and particularly, in recent years, of mothers of very young children – represents one of the most striking social trends of the late twentieth century in North America. This trend has many origins: the effects of the increasing urbanization of society on employment opportunities and on household needs, demographic factors, increased economic and financial pressures, and changes in attitudes toward work and family (Krashinsky, 1977; Kitchen, 1990). There is considerable evidence that many women are seeking employment as a response to economic insecurity: responding to high levels of inflation to maintain their families' standard of living; attempting to pull their families' income above the poverty line or toward some other level considered more adequate; dealing with the possibility of divorce, separation or widowhood; and providing for pension support during old age (Canadian Advisory Council on the Status of Women, 1984).

Several demographic and social factors point toward more Canadian women working outside the home in the future. On average, women are living longer, have much easier access to reliable birth control, and are having smaller families than they did a generation or two ago. Furthermore, the growth of jobs in the service sector, particularly those that employ large numbers of women, has been dominant in the second half of the twentieth century. Higher levels of educational attainment have also meant that more women are in a position to hold a wide range of jobs and to seek paid employment for career reasons rather than for monetary return alone. In general, attitudes toward "women's place" have been changing and, accordingly, the attitudes toward family and child nurturing have changed (Krashinsky, 1977; Ontario Ministry of Community and Social Services, 1987).

Canada has no national child care legislation or policy. Child care, like other health, education and social programs in the country, is under the jurisdiction of the provinces and territories. At the same time, maternity leave and parental leave policies are necessary adjuncts to a child care system. The federal government provides maternity and parental leave benefits under the Unemployment Insurance Act. In 1994, an eligible new mother could receive 57 per cent of her wages up to a maximum of $444, for a period of fifteen weeks. Following the fifteen weeks, either parent is entitled to parental leave benefits of the same amount for ten additional weeks. Adoptive parents are entitled only to parental leave benefits for ten weeks.

Individual provinces have their own legislation. All provinces and territories have maternity leave provisions and some have parental, adoption and paternity leaves. Only Ontario and Québec have modified provincial leave periods to allow parents to take advantage of federal benefits (Friendly *et al.*, 1991). For example, Ontario sets out the right of eighteen weeks unpaid parental leave, while Saskatchewan and the Yukon Territory do not have parental leave in their legislation. Québec legislation includes general child care leave for five days per year without pay to fulfil obligations relating to care, health, or education of one's minor children (Labour Canada, 1990).

In general, younger women have higher paid labor force participation rates than older women, undoubtedly because of their relatively higher educational levels and their expectations about the roles of women (see Statistics Canada, 1984). Table 3.1 shows that for mothers with children

Table 3.1 Paid labor force participation rates for mothers, Canada, 1975–1994

Age of youngest child	1975	1978	1981	1984	1987	1990	1994
Under 3	31.2	37.6	44.5	51.5	57.0	60.3	63.1
3–5	40.0	46.1	52.4	56.9	63.1	68.6	67.4
6–15	48.2	54.3	61.1	64.4	70.5	76.9	76.8
Under 6	34.8	41.0	47.5	53.6	59.4	63.6	64.8
Under 16	41.6	48.0	54.5	59.1	65.0	70.4	70.9
All families without children <16	42.3	44.9	47.3	49.1	50.6	53.8	54.1

Note: Participation rate represents the labor force expressed as a percentage of the population 15 years and older in that group. For each year, the participation rates are the annual averages; but for 1990 and 1994 the figures are for December.

Source: Statistics Canada Catalogue 71-001 (monthly)

under 6 years old, the paid labor force participation rate in Canada was already 35 per cent in 1975, and rose steadily to 65 per cent by 1994. The labor force participation rate for women with at least one child less than 3 years old is only slightly lower in each year, and was at 63 per cent in 1994 (these rates include women in both full-time and part-time employment). These trends show no signs of abating, and may also mean that there are fewer women at home to provide informal child care for others. Thus, the demands on the formal child care sector may increase faster than is implied by the increasing labor force participation rate of mothers. These figures are one means of illustrating why child care has become an issue of pressing social and political concern in Canada.

CHILD CARE PROVISION IN CANADA

As Friendly *et al.* (1991) point out, Canada does not have a child care *system*. The federal government has not taken a proactive or facilitative approach to developing such a system. Thus child care has developed with marked regional variations across Canada. The term "child care" refers to the care of children by someone other than their own parents or guardians.[1] There are four types of child care:

1 in-home care: where a child is cared for in his/her own home, by parent(s), other relatives or non-relatives such as a nanny or a sitter;
2 unlicensed family child care homes: in which a caregiver (usually a woman) takes care of one or more children (perhaps in addition to her own) in her home without regulation by public agencies (unregulated after-school programs, held in a variety of facilities, also exist);
3 licensed family child care homes: licensed by, and usually inspected by, the provincial government. These homes must meet standards of space, staffing, and other requirements set by each province (for example, a home may not care for more than five children); and
4 child care centers: licensed and inspected by the provincial government.

These same child care facilities could also be differentiated into formal and informal sectors; the formal sector is regulated by the government, while the informal sector is unregulated. The informal child care sector consists of types 1 and 2 above, while the formal sector consists of types 3 and 4. Presently, about 20 per cent of preschool children receiving child care in Canada are found in the formal sector. However, given the absence of data on the informal sector and that the focus in this chapter is on service provision by public agencies and regulated private sector firms, the present analysis only concerns the fourth type – licensed child care centers (see Dyck, this volume, for discussions of informal care).

GOVERNMENT INVOLVEMENT IN CHILD CARE

In this section I discuss the different levels of government's responsibility regarding child care. The province has responsibility for regulation and partial funding of child care. Municipal governments are responsible for the administration of funding. There are other forms of government involvement in child care, notably through the municipal governments' planning departments and the Boards of Education.

Provincial responsibility for regulation

The first creches and nurseries in Canada were established in the late nineteenth and early twentieth centuries and were usually run by charitable organizations. They freed poor women, often sole-support mothers, from their domestic obligations and made them available for menial and domestic jobs. A creche run by Roman Catholic nuns opened in Montréal in the 1850s; the first day nursery in Ontario was founded in 1891 by wealthy philanthropic women. Child care was dispensed as a charity. Following the passage of the Mothers' Allowance Act in 1920 (to provide income to single mothers with young children), the use and availability of child care declined; in particular, agencies ceased to provide infant care.[2] By 1933, there were twenty day nurseries in Canada, serving about 2,500 children. But as late as 1940 some single parents even placed their children in orphanages (where they could be visited on weekends) in order to work; there was no other suitable care available (Schulz, 1978; Goelman, 1992).

Direct government involvement in child care emerged during the Second World War, when women were needed in the labor force to work in war-related industries. Before this time, public opinion held that the only legitimate reason for a mother to go to work was dire financial necessity (Stapleford, 1976). Both Canada and the US enacted legislation to set up nurseries and provide child care. In Canada, a federal Order-in-Council set up the Wartime Day Nurseries Act in 1942. This Act split operating and capital costs evenly between provincial and federal governments. Ontario and Québec were the only two provinces to act under the agreement (Schulz, 1978), partly because farm work, which dominated in several of the other provinces, was not considered an essential war industry. Three-quarters of the spaces were allocated to children whose mothers worked in essential war industries. In 1946, the federally sponsored centers closed as the war ended, although public pressure led Ontario to keep some of the nineteen established centers open. The province enacted its first Day Nurseries Act in 1946 and assumed responsibility for child care centers on a 50/50 cost-sharing basis with municipalities (Women's Bureau, 1981). The Act also required the licensing and inspection of child care centers and nursery schools.[3]

39

Currently, formal child care in Canada is regulated by provincial governments, but the funding of subsidies is shared by different levels of government, as is common in many Canadian social programs. (Only in Ontario do third level governments, i.e. municipalities, pay 20 per cent of costs, while the federal government pays 50 per cent and the province 30 per cent. In all other provinces eligible costs for subsidies are shared 50/50 by federal/provincial governments.) In 1964, the Canada Assistance Plan (henceforth referred to as CAP) was passed by the federal government; and since 1966 it has provided for federal/provincial cost sharing of several welfare services, including child care. Under CAP, parents who meet particular criteria (income or need criteria) may qualify for assistance in meeting the cost of child care services. Under this federal/provincial cost-sharing scheme, a province may subsidize families "in need" or likely to be "in need" to help meet their child care expenses. A large proportion of these families are headed by lone parents. However, the federal funds are not available unless a province opts into this program. Since the late 1970s, all provinces have participated in the program.

Since CAP was introduced, the number of child care spaces in Canada has grown, at first slowly, then rapidly in the 1970s (see Chapter 1, footnote 2). The growth in formal child care spaces has been discussed elsewhere (Mackenzie and Truelove, 1993; Goelman, 1992); but there are presently major differences across the provinces (see Andre and Neave, 1992, for information intended for parents regarding the child care situation in every province and territory). The formal child care sector does not provide enough spaces for all those children who need it or for families without subsidies. By using four different estimates of children who need full-time care (ranging from children whose mothers are in the labor force, full-time or part-time, to only those children whose parents are working full-time), the *Status of Day Care in Canada* (Health and Welfare Canada, 1991) estimates that the proportion of care available in the formal sector is as follows:

- less than 14 per cent of infants (age 0 to 17 months) who require full-time care can be accommodated;
- less than 21 per cent of children 18 to 35 months;
- less than 49 per cent of children 3 to 5 years;
- and less than 8 per cent of children 6 to 12 years.

In short, Canadian formal child care best serves the 3- to 5-year-olds. There is a desperate shortage of formal care for infants (infant care also tends to be more expensive).

CAP involves two provisions, one for welfare services, the other for social assistance. The welfare services provisions of CAP allow for the cost-sharing of rehabilitation services, counseling, assessment and referral, adoption services, community development, homemaker services, and child care, provided to persons in need or likely to be in need if they do not receive the service. The philosophical basis for this part of CAP is to remove some of the causes of poverty and dependence on social assistance. It allows for the funding of costs to *agencies* providing the services, rather than to individuals or families receiving the service. The agencies must be provincially approved and cannot be commercial if the cost is to be shared by the federal government. In British Columbia and the Northwest Territories, which use the welfare services route, people using unregulated (informal) family child care can be subsidized.

In contrast, Ontario has long used the social assistance section of CAP and now so do Newfoundland, Prince Edward Island and New Brunswick. This section allows for the funding of direct financial payments to *persons* in need (general social assistance, commonly called "welfare," to pay for food, clothing and so on) and for the funding of four "prescribed services" that may be provided to a person in need: rehabilitation services, child care, homemaker services and counseling. Under CAP, persons are "in need" when they are unable to provide adequately for themselves or their dependents. Eligibility is based on a needs test established by the province (based on federal guidelines) that takes into account budgetary require-ments, income and resources available to meet those requirements. In Ontario, criteria for the "prescribed services" are not as stringent as for "general social assistance." Unregulated (informal) child care is eligible for funding under this section of CAP, but usually only formal child care costs are subsidized by provinces. However, over time the federal government has limited its funding contributions for child care. This has meant that, regardless of which section of CAP the provinces adopt, they either have to pay a higher share of child care costs, or decrease the number of subsidies provided (for a fuller explanation of CAP funding for child care, see Friendly, 1994 and Townson, 1985).

In Ontario, the provincial government's Ministry of Community and Social Services administers the federal government's 50 per cent share and the province's 30 per cent share of funding for subsidized child care under the terms of the Ontario Day Nurseries Act. Provinces are responsible for regulating formal child care, and the Act provides for the licensing of child care centers and sets standards for the provision of care. Both types of formal child care – centers and supervised private homes – are eligible to receive children with subsidies.

The CAP has had enormous influence on the development of provincial child care programs, the setting of fees and funding mechanisms. CAP was never intended to fund child care, although both sections of CAP make government funding of child care possible. It funds people or welfare services for selected families "in need" or likely to be in need. Beyond this, there is an income tax deduction for child care expenses: the Child Care Expense Deduction that all parents with receipts (for formal or informal child care) may use (see Friendly *et al.*, 1991: 19–20). Provincial and municipal levels of government can choose to focus on other criteria that meet CAP regulations. In Ontario, CAP has been interpreted as enabling lone parents to work; reducing dependence on public assistance; enriching the lives of low-income children; assisting ill parents; assisting the adaption of immigrants; and assisting children with disabilities (Ontario Ministry of Community and Social Services, 1981). However, in keeping with shifting public opinion, the ministry has recently emphasized the need for child care to evolve from a welfare service to an essential and "mainstream" public service (Beach *et al.*, 1993; Ontario Ministry of Community and Social Services, 1992).

Municipal responsibility for the administration of funding

Regulation of formal child care is a provincial responsibility, and if an Ontario municipality opts into the Canada Assistance Plan's 50/30/20 cost-sharing agreement, administration of the fund becomes a *municipal* responsibility. This section explores how Metropolitan Toronto operation-alizes this responsibility. Metropolitan Toronto is the largest urban area in Canada; its combined population is 2.3 million people in an area of 241 square miles. It was formed in 1953 and is composed of the City of Toronto and five adjacent municipalities: East York, Etobicoke, North York, Scarborough and York. While Metropolitan Toronto is a regional government, each municipality within it also has its own government. Some responsibilities are shared between the local governments and Metropolitan Toronto (for example, affordable housing), while others are the sole responsibility of each municipality (for example, local zoning and garbage pick-up).

Metropolitan Toronto, like other Ontario municipalities, develops policies and allocates funding for subsidized child care services through the Children's Services Division of the municipality's Community Services Department. The municipality develops its own guidelines within the framework of the province's "needs test" to determine a family's eligibility for subsidy. However, many observers consider the annual needs test (taking into account housing, transit and other costs) to be intrusive and degrading (Ontario Ministry of Community and Social Services, 1987). In addition, Metropolitan Toronto introduced a first-come, first-served policy for access to subsidized spaces in 1983. This raises an important equity question as it is quite possible that very needy families are lower on the waiting list than are other, less needy, families who applied for subsidies earlier.

Metropolitan Toronto continually plans, reviews and evaluates its provision of subsidized spaces, particularly as funding decreases; and the equity of the provision of subsidized spaces, across the municipality has recently been evaluated (Community Services Department, 1990, 1992) (see Skelton, this volume, for a discussion of the spatial inequities in child care centers in Ontario). In the past, Metropolitan Toronto and the provincial government negotiated a budget for subsidized child care services each year (Friendly and O'Neill, 1983). Metropolitan Toronto's requests for the province to expand subsidized services were not met in the late 1980s by the provincial government. In the 1990s, the province has been forced to pursue a policy of even tighter fiscal restraint, as the federal government has capped its share of the costs. Thus, Metropolitan Toronto has had a very difficult time negotiating for its desired budget level; in fact the number of subsidized spaces has actually been decreasing each year recently. Moreover, Metropolitan Toronto does not have the tax base to keep the same number of child care subsidies: this would raise Metropolitan Toronto's share of the cost well above 20 per cent.

Aside from funding, the Children's Services Division of Metropolitan Toronto's Community Services Department has also developed its own criteria for child care center program *quality* to be used to determine whether a program will be allowed to accept children with subsidies. When a child care center applies for approval to care for subsidized children, Community Services Department staff inspect the center. If approved, the center signs a "purchase-of-service agreement" with the Community Services Department, and Metropolitan Toronto pays the appropriate number of subsidies directly to the center each month (based on the number of parents who choose that center and on their children's attendance records). Until 1983, Metropolitan Toronto had a policy of quotas deciding the maximum number of children with subsidies allowed at each approved child care center. Now, a parent who obtains the subsidy can use the subsidy at any of the over 350 full-time centers with purchase-of-service agreements, provided there is space at that center. Unfortunately, there is a lengthy waiting list for subsidies. The Metropolitan Toronto's Children's Services Department produces an "Infosheet" on subsidy spaces. The waiting list for subsidies grew dramatically in 1992 and 1993. The July 1993 subsidy waiting list consisted of eligible low-income families waiting for 5,011 infant spaces, 2,795 toddler spaces and 6,309 preschool spaces.

Other forms of local government involvement

Local governments within Metropolitan Toronto also have some regulatory control over child care centers, mainly through their planning departments (local governments have control over local land-use zoning). In the past, some have developed zoning by-laws restricting child care programs to certain areas. However, these by-laws are no longer considered to be a problem: in all parts of Metropolitan Toronto a child care center is allowed in commercial districts, in buildings such as schools and churches in all residential neighborhoods, and in some residentially zoned areas. This change occurred gradually as the numbers of mothers working away from home rose and the demand for child care increased. There are some industrial districts in which a child care center still cannot open. According to zoning by-laws, home-based child care is allowed to operate in any residential district as long as five or fewer children are cared for; with more children a home would have to be licensed as a child care center. However, in most cases, the cost of building new centers or renovating existing structures to conform to the requirements of the Day Nurseries Act is actually more burdensome than local by-laws. For example, under the Act, child care centers may not be located above the second floor of a building and must have direct access to a ground level outdoor play area.

Some local governments have moved beyond their narrow regulatory control of centers. For example, recognizing that child care workers are among the most poorly paid, the City of Toronto provides annual grants to non-profit child care centers to increase the salary and benefit levels of their employees. The City of Toronto also encourages the development of non-profit, workplace child care centers in negotiations and trade-offs with developers of downtown office towers (Beach *et al.*, 1993).

A further complication in location questions surrounding child care is that local Boards of Education also play a significant role (in Metropolitan Toronto each local municipality has its own Board of Education). Since 1987, the province of Ontario has had a policy of building a child care center in all new schools. Declining enrollment has made empty classrooms available, so space is leased to formal child care programs, with lower rents for non-profit centers than for commercial centers. Outside the City of Toronto, the YMCA is the main provider of school-age child care programs in school spaces in Metropolitan Toronto Boards of Education. Indeed, this researcher found that 37 per cent of all full-time child care centers in Metropolitan Toronto are located in elementary schools or high schools. When school-age only programs – most of which are located in schools – are added in, then 45 per cent of child care centers are located in schools. Some Boards of Education in Metropolitan Toronto provide start-up funds to these centers. Some will allow only non-profit and parent-run cooperative centers in their schools. The City of Toronto Board of Education, for example, has a goal of providing a non-profit child care center for most of their schools: where there is space and community need. The same Board also actively encourages and supports the formation of parent-run cooperative child care centers and has aggressively sought subsidies for its centers from Metropolitan Toronto. Boards of Education also lease space to supervised home child care agencies and to support services such as toy lending libraries and drop-in centers; such facilities help informal child care providers as well as parents. The Boards of Education also provide small child care centers of subsidized care for the children of teenagers completing high school, and Metropolitan Toronto provides subsidies for these spaces.

CONCLUSIONS

This chapter has outlined the regulation and funding of formal child care in Canada. Child care serves a vital role in permitting women to participate in the labor force. In the 1980s and 1990s, the paid labor force participation rates for mothers of young children have continued to increase; by 1994, 65 per cent of women, who were mothers of children under 6 years old, worked for pay. During this time period child care came to be perceived as an essential and "mainstream service" not just a welfare-oriented service. Regulation of child care is a provincial responsibility, while the allocation of funding is a municipal responsibility. Beyond this, local governments can be involved in shaping the form of child care provision in a variety of ways. These include land-use planning decisions and the introduction of progressive policies regarding the pay of child care workers. Moreover, Boards of Education also play a role. For instance, they may make space in schools available to certain types of child care for relatively low rents. However, there is now no new funding help from the provincial government to create new child care centers and it is unlikely that more will be located in schools in the near future.

The roles of each level of government alter over time in response to social, political and economic change. In the near future, financial factors may force municipal governments and Boards of Education to cut back their involvement in the provision of child care. The recent recession and the conservative political agenda are now influencing the federal and provincial governments; stringent funding cutbacks affecting child care mean that operationally the Canada Assistance Plan no longer functions. The 1990s have seen a decrease in the number of subsidized child care spaces, and continued decreases in spaces and closures of centers are expected.

Part III

THE PROVISION OF CHILD CARE AT THE STATE AND PROVINCIAL LEVEL

4

MAKING THE TRANSITION TO SCHOOL

Which communities provide full-day public kindergarten?

Ellen K. Cromley

KINDERGARTEN AS A FOCAL POINT

Across the United States in public policy debates on child care and other family support measures (such as maternity leave), the working parent's need for child care is seen as a problem primarily from the time the child is born until the child enters school (see Bloom and Steen, this volume, for a discussion of the need for more and better child care programs in the US). By the late 1980s, there were more children under age 5 than at any time since the mid-1960s (Trost, 1988), and currently 60 per cent of mothers of preschool children are in the paid labor force (US Bureau of the Census, 1992a). The availability and quality of child care for preschool children have, quite rightly, become important issues. However, 75 per cent of mothers of *school-age* children are also in the paid labor force (US Bureau of the Census, 1992a). And these working parents can attest to the new set of challenges posed by the child's entry into school, including arranging before- and after-school care, planning care for irregularly occurring school vacations, holidays, and early closing or staff development days. Beyond this they have to coordinate the child's increasingly complex and independent activity patterns with those of other family members.

As more mothers have entered the paid labor force in the US over the last two decades, there has been an increase in interest in early childhood – as distinct from elementary – education. This reflects a growing recognition that early childhood experiences have significant implications for later schooling (Rudolph and Cohen, 1984; Cryan *et al.*, 1992). Kindergarten, usually the first year of public schooling for the 5-year-old child in the US is the nexus of most issues associated with child care and early childhood education. The kindergarten year is often particularly difficult for working parents because many kindergarten programs are half-day, scheduled as 2½-hour sessions, five days per week (Cryan, *et al.*, 1992). Parents must arrange for separate before- or after-school care and for transportation

49

between school and the child care provider during the middle of the normal working day. Increased interest in full-day kindergarten, however, has raised issues about the kinds of educational activity appropriate in a full-day program (Fromberg, 1987). Most recently, the potential role of public schools as sites for child care provision has been vigorously debated.

Most educators agree that decisions about the content and length of an educational program should be made on the basis of what best serves children (Peck *et al.*, 1988; Walsh, 1989). Indeed, given that close to 90 per cent of all 5-year-olds in the US attend kindergarten (Sassower, 1982; Karweit, 1988), any change in the organization of kindergarten could have widespread consequences. Nevertheless, it is clear that the sometimes conflicting needs of parents, teachers and administrators influence programming. Until the Second World War, many kindergarten programs in the US were full-day (Connecticut Early Childhood Education Council, 1983). The teacher shortage during the war required conversion to half-day programs, although some child care centers were then funded through public schools. The choice for half-day programming was reinforced by cost and space considerations as the school-age population expanded rapidly during the 1950s (Puleo, 1988).

To the extent that conditions in the larger society affect the organization of child care and public schooling, it is worth examining how these are manifest at the local level, where the responsibility for the provision of elementary education lies in the US. The power to provide and regulate education is not explicitly mentioned in the US Constitution as a power of the federal government and, as a result, the provision of public education is left to state governments. Provisions for the establishment of public school systems date from the 1820s to 1850s when new constitutions were drafted in many states (Kelly and Harbison, 1970). In most states, provisions were made for the creation of local school districts to be funded through local property taxes. Although local control and funding of schools has been vigorously debated over the last several decades (because of the disparities in educational opportunities observed across districts), it remains the norm. There continues to be "great variation in goals and standards from state to state and district to district" (Rudolph and Cohen, 1984: 7), perhaps more than would be found in a situation where public education is administered by the central government. These arrangements mean that there is likely to be an association between the socioeconomic characteristics of a particular community in the US and the types of educational programs available within that community.

To investigate the possible connections between community characteristics and the organization of early childhood education and child care, this chapter considers the geographical patterns of availability of full-day kindergarten programs in the state of Connecticut. Connecticut is a relatively small state in terms of area and population size (see Table 4.1). It is also one

Table 4.1 Socioeconomic characteristics of the United States and Connecticut, 1990

Characteristic	United States	Connecticut
Population	248.7 million	3.3 million
Per cent Black	12.1	8.3
Per cent Hispanic	8.9	7.4
Median family income	$35,225	$49,199
Per cent of women with children under 6 in the labor force	59.7	60.8
Per cent of kindergarten programs that are full-day	33.6	34.1

Sources: US Bureau of the Census (1992a, 1992b, 1992c)

of the wealthiest states in the US in terms of median family income. The presence of minority populations and the proportion of women, with children under 6 years old, who are in the labor force are similar to the levels in the nation as a whole. The percentage of towns in Connecticut belonging to school districts where full- or extended-day programs are available is similar to the percentage of principals in a national sample of schools who reported full-day kindergarten programming, although full-day enrollment is a lower percentage of kindergarten enrollment in Connecticut than nationally (Gardner, 1986). Community characteristics reflecting a set of issues surrounding the value of full-day kindergarten (the educational issue, the environment-of-the-child issue, and the child care issue) are identified for each town in the state. Discriminant analysis provides a useful evaluation of how these characteristics distinguish towns belonging to school districts where full- or extended-day kindergarten is available from those towns belonging to school districts where it has not been available.

THE EARLIEST KINDERGARTENS

The first kindergarten was organized by Friedrich Froebel in Blankenburg, Germany, in 1837 as a *Kleinkinderbeschaftingungsanstalt*, literally an "institute for the occupation of small children" (Research Division, National Education Association, 1962). Froebel's contribution to early childhood education was the concept – then revolutionary – that play was an important means of physical and mental growth in young children. The methods of guiding children in play that Froebel espoused were later rejected by other educators as overly limited and formalized (Weber, 1969). These criticisms, however, were directed more at the effectiveness of his methods than at the soundness of his aims. The implications of Froebel's views on education, developed through years of teaching in private schools and tutoring students before the first kindergarten was opened, were

sufficiently far-reaching that the Prussian government prohibited kindergartens in the aftermath of 1848 (Lilley, 1967). By this time, however,
Froebel had established a circle of teachers and pupils who carried his
philosophy and methods to the US during the primary period of German
emigration in the late nineteenth century (Weber, 1969). Over the last 150
years, educational programs based on Froebel's principles have spread to
many other countries, including the US.

The first kindergarten in the US was opened in 1856 in Watertown,
Wisconsin, by Margaretha Meyer Schurz (Weber, 1969). Although her
program was organized for German-speaking children and was short-lived,
Schurz influenced the spread of kindergartens in the US through her
contacts with others. Indeed, "the largest number of kindergartens directed
by teachers trained in Germany was found along the east coast" of the US
(Weber, 1969: 21). The pioneering US kindergartens were started by
Germans who subscribed to Froebel's views and they were initially available
as full-day, private programs for parents who could afford them. Henry
Barnard, Secretary of the Connecticut Board of Education, introduced
Froebel to US educators in 1854 in a report to the Governor and in
subsequent articles in an education journal published in Hartford,
Connecticut (Weber, 1969). The first kindergarten in the US for English-
speaking children was opened by Elizabeth Palmer Peabody in Boston in
1860 (Research Division, National Education Association, 1962). Boston
established the first public school-based kindergarten in 1870, but it lasted
only a few years and another was not established until 1887. In 1873,
kindergarten became part of the public school system in St. Louis, Missouri.
This change exposed kindergarten programs to the forces playing upon the
public schools at large. It also brought into sharper relief the educational
issues of managing the transition between kindergarten and primary school,
and the educational activities appropriate to each.

A major impetus for the diffusion of kindergartens within the US was the
social reform movement of the 1880s and 1890s (Weber, 1969).
Kindergartens were organized in settlement houses as a matter of
philanthropy to advance the well-being of young children. The roots of
the role of the school in undertaking what was not being accomplished in
the home – the environment-of-the-child issue – is the extension of
kindergarten to underprivileged and minority populations. The environment-
of-the-child was also raised in the 1930s when kindergartens were
evaluated, along with nursery schools and child care facilities, as part of a
White House conference on Child Health and Protection. A central issue
was that many children were "handicapped" by the home environment in
which modern urban society forced them to live (Committee on the Infant
and Preschool Child, 1931). In the same vein, the increase in the number of
wage-earning mothers was also considered a social and economic change
warranting public attention.

By the late 1950s, 70 per cent of all urban places in the US had public kindergartens, although kindergarten was mandatory in only six states – California, Illinois, North Dakota, Oregon, Texas and Utah (Steiner, 1957). According to data from a questionnaire completed by a sample of kindergarten teachers in 1961, only 3.5 per cent of public kindergartens had full-day sessions. Major changes have taken place in kindergarten programs in the US since then.

KINDERGARTEN TODAY AND THE FULL-DAY ISSUE

By the early 1990s, only four states (Arkansas, Delaware, South Carolina and Virginia) had compulsory school attendance requirements for 5-year-old children (National Center for Education Statistics, 1993). Some other states require school systems to provide kindergarten, but do not require attendance. A number of states have statutes permitting kindergarten, but mandating neither provision nor attendance. Nationally only 13 per cent of kindergarten students were in full-day programs in 1970; by 1991, 40 per cent attended full-day sessions (National Center for Education Statistics, 1993).

The growth of full-day programs has been geographically uneven, with more impressive expansion in certain states and regions of the country. For example, Hawaii has had full-day kindergarten in every elementary school since 1955, while in New York state, the majority of students were in full-day programs by the 1985–1986 school year (Puleo, 1988). On a regional basis, a high percentage of full-day programs are found in the southeastern US (Gardner, 1986). A possible explanation for this is that the full-day programs may be provided for enrichment, as a relatively low percentage of kindergarten pupils arrive at school with a full year of pre-kindergarten experience. In the northeastern states, by contrast, more than half of the pupils have had at least a year of preschool experience.

A number of factors have contributed to the resurgence of interest in a full-day kindergarten session. One of the most obvious reasons for reconsidering full-day programs has been their potential for improving basic academic skills (the educational issue). Research on the impact of kindergarten session length generally shows better outcomes for full-day programs, although the use of different methodologies limits the extent to which results from the various studies can be compared (Puleo, 1988; Cryan et al., 1992). The benefits appear particularly great for children of low socioeconomic status (Peck et al., 1988).

The compensatory value of a longer kindergarten session possibly arose in some US educators' minds because of the success of the Head Start program, a program of early childhood education for disadvantaged children that was instituted in the 1960s as part of the War on Poverty (Lazar et al., 1982). From the point of view of administrators, full-day programs, which, according to one study, increase costs by 20 to 24 per cent (Puleo, 1988), may be considered an appropriate use of resources if they contribute to grade retention and reduce referrals for special education or remedial education later on. It has also been suggested that full-day programs generate savings by reducing transportation costs. However, savings have also been realized merely by scheduling full-day sessions on alternate days (Ulrey *et al.*, 1982). Several studies have supported full-day kindergarten as a means of enhancing enrollment (Puleo, 1988). Apparently, some working parents who prefer full-day programming, enroll their children in private full-day programs if half-day programs are the only option in the public schools, leading some of these parents to keep their children in private schools beyond kindergarten. Parents, teachers and students involved in full-day programs are generally very enthusiastic (Peck *et al.*, 1988), while parents not directly involved in full-day kindergarten programs are less supportive. Concerns for their child's ability to cope with a full-day program and desires to maintain contact with their child in the home are the reasons most frequently given.

Issues surrounding the appropriate content of a full-day program are complex. There is currently tremendous downward pressure in the primary curriculum, including calls for public school instruction of 4-year-olds (Hiebert, 1988). At the same time, others are speaking to the perils of early academic demands (Karweit, 1988). Of course, because being in school for a full day does not necessarily require that children be involved in academic work for the whole day, this debate is not likely to be resolved by setting the length of the kindergarten day alone.

CONNECTICUT AS A SETTING FOR THE STUDY OF FULL-DAY KINDERGARTEN

The urban areas of Connecticut extend along the coastline from the New York border to New Haven and then up the central valley into the Hartford metropolitan area (see Figure 4.1 for locations). Historically, towns in this urban corridor, including Greenwich (part of the New York City commuting zone, like other towns in the southwestern part of the state) and Hartford, played important roles in the development and adoption of early childhood programs in Connecticut and the US. Public education is a function of town government, the basic unit of local government in New England, unlike other regions of the US. All of the territory of the state lies within the boundary of one municipality meaning that there are no unincorporated areas. There are 169 towns in Connecticut, but not every town functions as its own school district. In terms of public education, Connecticut's towns fall into three groups. One set operate kindergarten to grade twelve education, and they tend to be in urban areas. The second set offer elementary education and varying levels of secondary education, but not the complete range. The third group do not operate their own schools, they tend to be smaller towns (often with populations of less than 2,500) located in the rural northwestern and northeastern regions of the state. The levels of education not offered by towns in the second and third groups are dealt with by sending students to schools run by another – not necessarily neighboring – town, or to a school in one of the "regional school districts" that have multi-county catchment areas (in these cases the towns pay a per pupil tuition to the town or district operating the school). In all there are 166 local public school districts in the state, 149 are town-operated and seventeen are regional districts (State Department of Education, 1992).

Connecticut state law requires the provision of kindergarten, but attendance is not mandated. The minimum age at which children are required to be enrolled in school is 7. Data on private kindergarten programs in the state were not available. Public kindergarten enrollment for the 1990/1991 school year was close to 40,500 (State Department of Education, 1992). Seventy-eight per cent of students attended half-day, 6 per cent attended extended-day (which in Connecticut means a session length somewhere between 2½ hours per day and full-day), and 16 per cent attended full-day (a session comparable in length to the regular elementary education session). In nineteen towns, all kindergarten enrollment in 1990/1991 was accounted for by extended-day or full-day enrollment. In every other town having full-day or extended-day enrollment (twenty-four towns), that enrollment represented only a portion of total kindergarten enrollment.

Recent efforts to consider the value of full-day kindergarten programs in Connecticut schools date back to 1979, when a task force was established by the Connecticut Early Childhood Education Council, an advisory council to the State Department of Education (Connecticut Early Childhood Education Council, 1983). The task force included members of the Kindergarten Association of Connecticut and the Elementary and Middle School Principals Association of Connecticut. The task force intentionally focused on full-day kindergarten in the public schools. The report of the task force was published in 1983. Its purpose was to provide information to enable individual schools and communities to study the issue and make their own programming decisions. At that time, the continuous 2½-hour session counted as a full-day equivalent for calculating the number of students enrolled in a public school, one of the factors taken into account in determining state aid to local districts. Lengthening the session would not increase the number of students or the amount of funding expected from the state. To support the development of full-day kindergarten programs, a grant program was established in 1987 but was terminated at the end of the 1990/1991 school year as part of cutbacks made to resolve the state budget crisis. Nevertheless, twenty-nine towns had grants supporting full-day kindergarten for at least one year during the period that the program was in operation (Lester, 1990). The pattern of availability of full-day and extended-day programs, and the issues surrounding implementation of full-day and extended-day kindergarten warrant investigation of the community characteristics associated with the presence or absence of these programs.

MODELING THE AVAILABILITY AND DIFFUSION OF FULL-DAY KINDERGARTEN

Geographers have paid relatively little attention to the organization of care and education in early childhood. A few studies on community location of child care programs are exceptions (Cromley, 1987). As an extension of that work, this chapter considers connections between community character-istics and the organization of early childhood education and child care. This case study and others in this book expand our understanding of child care, especially in terms of how it is provided in different geographical settings. One of the few early studies of particular relevance to this chapter is Meyer's (1975) analysis of the diffusion of Montessori education in the US. Montessori methods derive from an educational program that Maria Montessori first set up for the children of working mothers in a tenement improvement project in Rome in 1907 (Weber, 1969). Although her ideas generated considerable interest in the US initially, influential educators concluded that her methods were not compatible with US educational theory (Meyer, 1975). However, Montessori education was reintroduced in the US in the late 1950s through the work of Nancy McCormick Rambusch who modified the original methods developed in Italy and established a Montessori school in Greenwich, Connecticut, that became a center of origin for the diffusion of American Montessori in the US in the 1960s (Meyer, 1975).

Meyer's work considers Montessori education, available primarily at private preschools, as an innovation and examines the characteristics of *individuals* adopting this innovation for their children, as well as the temporal aspects of the diffusion process in the US. Using data for 1962–1972, she found that adopters tended to be middle- to upper middle-class residents of metropolitan areas, living in communities with a slightly higher than average proportion of college-educated residents. Meyer's analysis provides support for the investigation of *community* characteristics associated with the availability of full-day kindergarten.

This section looks at the relationship between the availability of full-day kindergarten and some pertinent characteristics of Connecticut's towns. The case study employs discriminant analysis, a statistical technique that uses a set of variables to distinguish between two or more groups of cases (see Clark and Hosking, 1986). In this instance, an attempt is made to distinguish between two groups of towns in Connecticut: (1) those towns belonging to school districts where public school students were enrolled in full- or extended-day kindergarten during the 1990/1991 school year; and (2) those towns where no public school students were enrolled in full- or extended-day kindergarten during the 1990/1991 school year (State Department of Education, 1992). The objective is to determine how well these two groups of towns can be distinguished from one another based on community characteristics that relate to the need for full-day kindergarten.

For each town, 1990 data were obtained for five community characteristics:

1 women in the paid labor force who were 16 years or older and had children under age 6, as a percentage of all women 16 or older with children under age 6;
2 median family income;
3 community type;
4 the availability of regular preschool programming in the town's school district; and
5 the school district type.

Variables 1, 2 and 3 were collected from the 1990 US Census (US Bureau of the Census, 1992b, 1992c) and variables 4 and 5 are from Connecticut's State Department of Education (1992). The last three variables are expressed as binary (0/1) variables. "Community type" differentiates urban from rural towns, with "1" representing towns with populations greater than 25,000 and "0" representing all other towns. The towns were also encoded as "1" if they provided regular preschool programming, and "0" if not. And "school district type" divided towns according to whether their school districts provided high school level education ("1"), or not ("0").

The variable choice reflects previous research. To the extent that full-day kindergarten serves a child care need, a positive association is expected between full-day kindergarten availability and the labor force participation rate of mothers with children under 6 years old (variable 1). To the extent that full-day kindergarten is now seen as an innovation enhancing children's later education attainment, full-day kindergarten would be more likely to be available in higher-income, metropolitan communities which, extending Meyer's research, are more likely to adopt early childhood programs (variables 2 and 3). To the extent that full-day kindergarten addresses the environment-of-the-child, public school systems offering a broad range of programs, including regular preschool programs through high schools might be more likely to offer full- or extended-day kindergarten as an option (variables 4 and 5). Full-day kindergarten directed at filling this need would also be more likely in low-income communities with large minority populations where parents do not have the financial resources to provide enrichment activities for their children in the home or other settings. Although the presence of children from disadvantaged minority populations might suggest the need for full-day kindergarten, a variable representing race and ethnicity was not included in the model. Connecticut, like many other places in the US, is characterized by a high degree of residential segregation by race and class. Close to 40 per cent of the Black population of the state, for example, reside in either Hartford or New Haven. Given that previous research such as Meyer's indicates that the environment-of-the-child view suggests that large, more urban towns have a higher availability of full-day kindergarten programming, then the minority populations are located in towns where school districts are more likely to offer full-day programs.

ASSOCIATIONS AMONG COMMUNITY CHARACTERISTICS AND FULL-DAY KINDERGARTEN

The discriminant analysis produced a set of statistically significant coefficients (see Appendix I for details). In terms of the variables that are best able to distinguish between towns offering full-day kindergarten and those that do not, urban community type and high median family income are very powerful. School district type and the availability of preschool programming in the school system contributed much less. Surprisingly, the labor force participation rate of mothers of young children was also of low magnitude. In fact, contrary to expectation, the model suggested that low rates of maternal employment are associated with the availability of full-day kindergarten.

The coefficients derived from the analysis were then used to classify each of the towns. The classification of towns that the discriminant analysis predicted based on the community characteristics was compared to whether or not full-day kindergarten was actually available in each town. The results of this exercise are reported in Table 4.2 and are mapped to show the spatial patterns (Figure 4.1). Table 4.2 shows that based on the five community characteristics, the discriminant analysis correctly classified 73 per cent (124) of towns according to whether or not they offered full-day programs (ninety-six towns are correctly classified as having half-day programs, and twenty-eight towns are correctly classified as having full-day programs).

Figure 4.1 indicates that the towns that actually have full-day

Table 4.2 Actual versus predicted full-day kindergarten

Actual group membership	Number of cases	Predicted group membership	
		Half-day	Full-day
Half-day	126	96	30
Full-day	43	15	28

'Grouped' cases correctly classified: 73.4%

Source: State Department of Education (1992); US Bureau of the Census (1992b, 1992c)

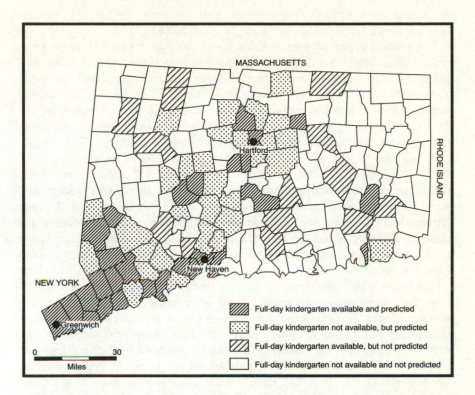

Figure 4.1 Geographical patterns of full-day kindergarten availability in
Connecticut, 1990

kindergarten programs available and that were correctly classified according
to the model (Group 1) are clustered in the more densely populated
southwestern part of the state and in the urban areas around Hartford in the
center of the state. These towns were all in districts offering kindergarten
through high school classes and almost all of them were urban communities
as defined in this analysis. Of the four groups of towns shown in Figure 4.1,
this group had the highest average median family income, the highest
proportion of districts where preschool programming was offered in the
school system, but the lowest average proportion of working women with
young children. Moreover, given that Meyer (1975) found a concentration of
Montessori programs in the southwestern region of the state, perhaps the
geographical pattern of Group 1 reflects diffusion from Greenwich as an
early center of innovation in early childhood education. At the same time,
the availability of full-day programs in most large urban towns in the state

means that disadvantaged minorities are also likely to be living in places where full-day kindergarten programs are available.

The second group of communities are those that the model incorrectly predicted as offering full-day programs and are differentiated from the towns correctly classified as offering full-day programs primarily by their community type. These towns were more likely to be smaller, suburban communities within the urban corridor stretching from Greenwich to Hartford. The third group of towns do have full-day kindergarten programs but were misclassified (not predicted) by the model. This group of towns, unlike Group 1, had the highest average proportion of working women with young children, but lower levels of median family income and availability of preschool programs. Most of the towns are scattered in the rural northwestern part of the state. Thus, the presence of full- and extended-day kindergarten in these towns reflects the national pattern of higher availability of full-day kindergarten in rural areas because their low densities and high transportation costs make full-day programming more feasible than other options.

The other rural communities, particularly in northeastern Connecticut, form the fourth group of towns. They were correctly classified as not offering full-day programming, and remain a core area of resistance to the adoption of full-day kindergarten programs. This group of towns were the least likely to be in school districts offering high school instruction, and the least likely to be in districts offering preschool programs. In 1990, towns in northeastern Connecticut had some of the highest percentages of young people of all towns in the state. However, many of these communities also have total populations of less than 2,000 distributed across rural areas with very little commercial development capable of supporting public education programs through property taxes. This pattern is consistent with Meyer's finding that low-income, rural residents were not adopters of innovative preschool programs.

DISCUSSION

The development of kindergarten programs in the US draws together a number of themes relevant to the question "Who will mind the baby?" In countries like the US, where education and many other social services are provided primarily at the state and local level, there can be significant variation in the availability of child care. Generally, high-income urban communities are more likely to offer a range of educational programs and services than low-income rural communities. Federal programs like Head Start have been important in low-income and minority communities, but the allocation of these resources is constrained by Congressional appropriation and also by the willingness and ability of localities to seek these funds, despite the evidence that early childhood programs are beneficial, at least during the first few years of elementary education. Localities are increasingly resisting federal and state mandates to provide certain types of educational and social services, unless they are accompanied by financial resources to provide those services.

Part of the variation in service availability also reflects the geographical patterns of diffusion of child care and early childhood education programs. As the history of kindergarten in the US illustrates, parents and professionals interested in the early childhood years have been willing to innovate and to consider the child care practices of other cultures, adapting them to local needs and possibilities. This point clearly underscores the need for more analysis of the nature and success of caregiving in various places, a key theme of this book.

Finally, based on this analysis of full-day kindergarten, it is worth considering whether the question "Who will mind the baby?" has been framed too narrowly. In the public policy debate, the child care issue is seen primarily as an employment issue. The question is interpreted as "*Who* will mind the baby?" From that perspective, the inverse relationship found between Connecticut mothers' paid labor force participation and the availability of full-day kindergarten would be the most important result of the analysis. Standing that result on its head, however, it is also worth asking why full-day programs seem to have taken hold in communities where there is apparently less need for them in the sense that parents have more time available to provide care themselves. The research presented here indicates that early childhood programs outside the home – even for children who have parents at home – are considered valuable in many communities because they contribute to the quality of life for children. In these communities, perhaps the central question of this book is interpreted as "Who will *mind* the *baby?*"

It can be argued that US public policy regarding the nature and availability of child care programs for infants, toddlers and school-age children will be most successful if it is based on solid research on the kinds of care that offer the best environments for children. It cannot be assumed that the availability of child care alone, regardless of quality, will solve the problems faced by working mothers. At the same time, evidence that appropriate childhood experiences outside the home are beneficial for all children, regardless of whether or not their parents work, might broaden the base of support for child care programs that would particularly benefit working mothers. More attention to child care patterns in different communities and their effectiveness will provide additional insight into the relationships between parental needs, community characteristics, and the care and education of young children.

5

CHILD CARE SERVICES IN ONTARIO
Service availability in a decentralized provision system

Ian Skelton

INTRODUCTION

The provision of licensed child care services in Ontario has expanded dramatically in recent years. In the last decade the number of child care centers in the province rose by more than 60 per cent to over 3,100 facilities. Nevertheless, it is widely recognized that the cost of quality child care is prohibitive for many households, that demand outstrips supply, and that there are substantial spatial differentials in service levels. Thus a persistent social policy issue is the availability of good quality child care. The central theme of this chapter is that increasing equity in geographic access to child care is an important planning goal. It is argued that the decentralized form of service delivery – that is, the form in which many services in the welfare state are manifest in Ontario – militates against attempts to foster equity in access to most personal social services, including child care. The chapter opens with a brief historical account of the development of the social service system in Ontario, showing that, in contrast with provision in other jurisdictions, services in Ontario have traditionally been organized on a decentralized basis. I then examine certain key characteristics of this form of delivery of social services. Next the development of the formal child care system in Ontario is described and then trends in the levels and geographic distribution of service are assessed. Finally, I discuss the potential contribution of recent policy shifts.

In the face of affordability and availability problems, many households use alternate arrangements, such as informal networks (see England and Dyck, this volume for a discussion of informal child care). However, this chapter examines formal child care through licensed center-based provision. There are two broad reasons for this focus. First, it is recognized that formal child care provides high quality service (Ontario Ministry of Community and Social Services, 1991). High standards are maintained because all center-based child care operators, whether they are munici-palities, voluntary sector agencies or for-profit corporations, must undergo regulation and inspection by the government. When center-based child care is seen as a valuable resource, its distribution is a matter of significance. Second, formal child care is an important point of state involvement in child care service, not only through regulation but also through subsidies in terms of support for user fees and other transfers. While limited support for informal care is extended through child care resource centers, for example, the greatest thrust of the response of the state to the social need for child care service can be examined through formal child care.

THE DECENTRALIZED WELFARE STATE IN ONTARIO

A marked trend toward decentralization of service planning and delivery has been identified in many western societies. In the US, control over the social service system was distributed to state and county administrations between the mid-1970s to the early 1980s, and provision was realigned into purchase-of-sevice arrangements with private agencies (Sosin, 1990). In Britain, privatization has taken place in social services as well as in many other areas (Le Grand and Robinson, 1984). These trends imply crucial structural changes in the welfare state in these countries. However, in the Ontario case, service provision has traditionally been the domain of private agents and planning and organization have taken place at the local level. In many ways the welfare state in Ontario has not decentralized, but has developed on a decentralized basis.

The federal role in social services has traditionally been in funding. In 1966, the Canada Assistance Plan (CAP) provided federal financing to provinces for social services on a cost-sharing basis. Federal efforts during the 1970s to assert greater control were hampered by recessionary pressures and constitutional conflict. The more recent federal attempts to restructure the welfare state by integrating services with job creation (Ismael, 1987) and with unemployment programs may have more impact on the nature of the services.

In Ontario the role of the provincial government has become established as that of a funder (with federal assistance) and regulator of services rather than a central provider. Over the post-Second World War period services have expanded very rapidly (Lang, 1974) and the provincial role has been examined carefully during this time. The Committee on Government Productivity (COGP), commissioned in 1969, advocated the "significant delegation of a substantial degree of responsibility for program delivery to agencies outside the government" (Ontario COGP, 1973). This view has been reflected within the Ministry of Community and Social Services (MCSS). The Task Force on Community and Social Services (the Hanson

Committee, Ontario TFCSS, 1974) recommended that the province should retain a leading role in central program planning and monitoring services, and that it should be involved in delivery only as a last resort. More recently the Social Assistance Review Committee (SARC) has attempted to shift the emphasis of the welfare state in Ontario toward services designed to reduce impediments to employment, a theme reflected in the recent policy initiative, *Turning Point* (Ontario MCSS, 1993a). Throughout these episodes of policy reformulation, the central–local division of labor has remained unchallenged. As was recently concluded: "We support retaining the present variety of delivery agents for the broad range of social and employment services available in Ontario" (Ontario SARC, 1988: 410).

Decentralization has been a consistent feature of the welfare state in Ontario. As Hurl (1984) has noted, the era of privatization brought nothing new to service users in the province. During the early 1980s it did appear that commercialization of social services might take place (Social Planning Council of Metropolitan Toronto, 1984; see also Mishra *et al.*, 1988). The Royal Commission on the Economic Union and Development Prospects for Canada (1985; the MacDonald Commission) proposed a mix of proprietary and voluntary service delivery. However, it would appear that in several service fields, particularly seniors' services and child care, explicit provincial policy is effecting a shift away from commercial delivery and that the mainstay of delivery will be the voluntary sector.

It is often suggested that there are benefits to decentralization. It fosters activity at the local level and, as Kramer (1981) points out, many useful formal programs began as local initiatives. However, in other instances the expected benefits of decentralization may not accrue. For example, Hurl and Tucker (1986) and Sosin (1990) examine the claims that decentralization will provide flexibility, responsiveness and efficiency to a service system. A decentralized system might be less flexible than expected if bureaucratic or political factors at the local level impede change. If providers become specialized then they might not act in a generalist way, undermining the responsiveness of the local system to the full range of needs. Finally, the economic efficiency often associated with decentralization may not materialize, and competition might be detrimental to relations among providers. In fact fewer resources may be available to the service sector if lobbying efforts are dissipated.

Decentralization to the voluntary sector poses a dilemma in terms of citizen participation. As Wolch (1990) points out, the consequences may be contradictory; on the one hand we have the possibility of expanded local control; and on the other, there may be increased state penetration of daily life. The effect is complicated by the gender division of labor within the welfare state (Williams, 1989). The caring role of women is emphasized as "women make up the majority of the salaried and volunteer workers involved in caregiving" (Guberman, 1990: 77). For example, in Ontario, the ratio of women to men on the boards of non-profit child care centers was recently estimated to be seven to one (Ontario MCSS, 1993b: 23).

There are other problematic aspects of decentralized service systems. Carroll (1989) discusses the need for accountability. Access to benefits and services under decentralization might depend on local decisions about who is deserving and who is eligible, rather than on broad social conventions under which state programs were established (Salamon, 1987). The principal concern of this chapter is geographical access to child care services. In a decentralized system, services will not materialize in a locality if no one takes the initiative to organize them – or if initiatives are unsuccessful. This clearly imposes hardship on people who need services since it assumes that needs are matched by the capacity to organize services and manage delivery. It also poses problems for planners attempting to equalize access because to attempt to create services directly would contradict the nature of the decentralized system.

DEVELOPMENT OF FORMAL CHILD CARE IN ONTARIO

Facilities for the care of children outside their homes have existed in Ontario since the second half of the nineteenth century (for the early history see McIntyre, 1979: Ch. 1; Stapleford, 1976). Until the Second World War, however, the number of centers was very small in relation to needs (Baker, 1987) and their purpose, for the most part, was to provide shelter and food for children in poor, fatherless families thus enabling the mothers to work (McIntyre, 1979), or to reduce infant mortality in other circumstances (Ledoux, 1987). In 1943, the province entered into a cost-sharing agreement with the federal government to subsidize care for children whose mothers assisted the war effort. This support was short-lived; in 1946 the federal government withdrew financial support. Many of the facilities established under the wartime program – about seventy centers serving some 4,200 children – closed in that year (Stapleford, 1976). Nevertheless, the program had important organizational and political consequences. The establishment of the Child Care Branch (originally the Day Nurseries Branch) within the ministry provided an organizational vehicle through which further child care programs could be promoted. In the political sphere, the wartime program helped to shift public opinion from a vision of child care as a charity to that of a necessary service (Krashinsky, 1977). Public resistance to child care center closures prompted the province to create the Day Nurseries Act of 1946 which facilitated ongoing provincial intervention in child care (McIntyre, 1979).

Public involvement in child care during the war set the stage for post-war policy development. However, because of the post-war closure of many of the child care facilities, in describing the expansion of services I consider the period from 1947 – the post-war low in child care provision – up to the early 1990s. This system development, of course, took place in the context of broader social changes and shifting gender relations, influencing the need and demand for service as women's participation in the labor force grew and household structure changed (Baker, 1987; Fooks, 1987; Gardner, 1984; Krashinsky, 1977; Ledoux, 1987; Ontario Advisory Committee on Day Care, 1976; Ontario MCSS, 1981, 1984; Ontario Select Committee on Health, 1987).

Since the Second World War, the province has attempted to make more funds available on an ongoing basis, by increasing its share of costs, by extending the types of agencies it was willing to fund and by extending the age range of children eligible. Financial assistance has generally been available only to children who were construed to be entitled on the grounds that they (or their families) required social assistance; ongoing general funding is a very recent innovation. It was as recently as 1988 that Direct Operating Grants were provided to the municipal and voluntary sector on a per child basis (Ontario MCSS, 1987). The grants were made available to commercial operators as well, but only those facilities providing services as of December 1987 were eligible and only half of the subsidy was payable. The other half would have come from the federal government through the Canada Assistance Plan (CAP), the cost-sharing program for social services, but commercial organizations are not eligible under this program. Most recently, Provider Enhancement Grants have been made available to raise the wages of child care workers, and a program of shifting proprietary agencies into the non-profit sector has been initiated.

Figure 5.1 shows the growth in child care services over the post-war period. The measure used is the number of facilities. This is a crude indicator, of course, because there can be large differences between centers in the amount of service provided. Data on child care center capacity can compensate for this limitation; however, for longitudinal study the data must be treated with caution, because standards for licensed capacity have changed over time. For example, between 1974 and 1975, the number of facilities increased by close to 100, but the licensed capacity dropped by over 3,200 places, about 7 per cent of the stock (Ontario Ministry of Treasury and Economics, 1978). The Provincial Secretary for Social Development had previously inflated the capacity figures artificially by dropping standards. Figure 5.1 shows that from the low point of only sixty centers in 1947 the supply grew to 3,137 in 1993. In the years immediately following the war, supply recovered somewhat, and it stepped upwards again in the late 1950s. The supply was maintained at a fairly steady level up until 1965. Growth since 1965 has been rapid, and while broader social changes clearly have driven the development of services, Figure 5.1 does suggest a specific influence of CAP funding in the latter half of the 1960s. Since the mid-1980s growth in the number of centers has accelerated as indicated by the steepness of the slope of the curve after 1984.

Number of Child Care Centers

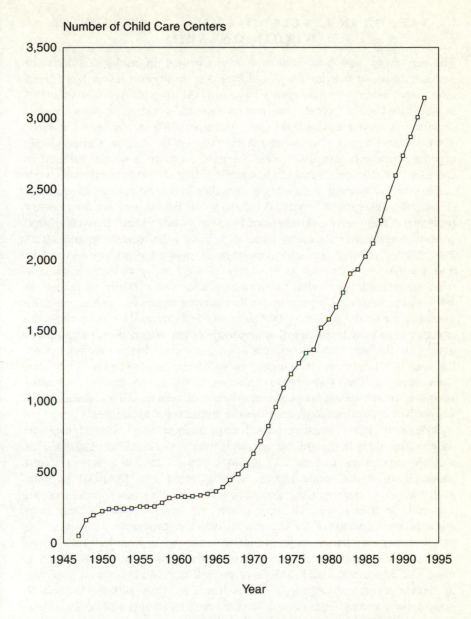

Figure 5.1 Number of child care centers in Ontario, 1947–1993

TRENDS IN LEVELS OF CHILD CARE SERVICE WITHIN ONTARIO

The very rapid growth in child care services since the mid-1960s does not necessarily mean that geographic differences in service levels have been eliminated. Several commentators have noticed unexplained discrepancies in the provision of child care service among geographic areas within Ontario (for example: Gardner, 1984; McIntyre, 1979; Ontario MCSS, 1981, 1984; Ontario Select Committee on Health, 1987); and, as Carter (1987) commented, the inequitable access across the province was mentioned in the SARC submissions. It has been suggested that municipal discretion, with respect to contributing to child care subsidies, has produced an inequitable distribution and gaps in services (Ontario MCSS 1992a). Indeed, the ministry recognizes that service levels may vary regionally and between large conurbations and other areas, especially if they are rural (Ontario MCSS 1981, 1984). Although at the provincial level, policy has addressed funding and regulation, it has not consistently tackled the problem of spatially uneven service levels within the province. One aim of a capital project in 1974 was to alter supply patterns so that service would be "made generally available across the province" (Ontario MCSS, 1981: 9). However, there do not appear to have been any planning tools or implementation mechanisms to support this aim. Some work was apparently done in the ministry during the mid-1970s on how to compare need among areas (Ontario Advisory Committee on Day Care, 1976), although, again, no procedures were adopted on an ongoing basis. For child care, as well as other services, local supply has depended exclusively on the initiative of local agents.

A current policy initiative, Child Care Reform, explicitly attempts to redress inequities in geographic access (Ontario MCSS, 1992a, 1992b). This strategy recognizes the complexity involved in initiating services, and proposes to make training and administrative and financial support available to under-supplied geographic areas. This may represent an important shift in service delivery policy, to the extent that it may help equalize the capacity of local areas to develop programs. However, the policy shares its reliance on local initiative with what went before.

The remainder of this chapter addresses these issues in two ways. First, there will be a longitudinal analysis of the differential impact in the growth in service levels for Ontario counties between 1954 and 1993. Second, geographic patterns in the current service levels by county will be identified. Ontario's geography is expected to have an effect: the northern region is generally sparsely populated, while the southern portion is highly urbanized, forming part of the Windsor–Québec corridor, containing Toronto (the economic and political center of the province) and Ottawa (the nation's capital).

Table 5.1 Child care centers per 1,000 children under age 5, Ontario counties, 1954–1993

	1954	*1964*	*1974*	*1984*	*1993*
Mean	0.30	0.36	1.53	2.90	4.02
Std Dev.	0.32	0.29	0.74	1.14	1.45
Std Dev./Mean	1.07	0.82	0.48	0.39	0.36
Minimum	0.00	0.00	0.26	1.01	1.46
Maximum	1.25	1.05	3.25	6.52	7.86
Gini	30.56	22.99	12.63	12.67	10.21

Source: Canada Dominion Bureau of Statistics (1953, 1962); Ontario Ministry of Community and Social Services (1993c); Statistics Canada (1976, 1981, 1992)

Table 5.1 reports the trends in service levels within the province between 1954 and 1993. The boundaries of the counties have changed over time and in order to account for these changes a consistent set of forty-seven counties was constructed. Table 5.1 presents descriptive statistics associated with the number of centers per 1,000 children under 5 years old. It also reports the Gini Coefficient, which is an index of dispersal and concentration of facilities relative to the total population (Smith, 1977). The Gini Coefficient ranges from a minimum of zero (if facilities are distributed in exact proportion to population) to a hypothetical maximum of 100 (if facilities are all concentrated at one location).

Table 5.1 shows that the mean ratio of child care centers per 1,000 children under 5 years old has grown considerably over the period 1954 to 1993. During the first decade, change was relatively slow; the average increased from 0.30 to 0.36 child care centers per 1,000 children aged under 5 years old. Between 1964 and 1974, the service ratio increased more than fourfold to 1.53 and by 1993 it had reached 4.02. On its own, the standard deviation seems to indicate that the variance has also increased over time, suggesting that there is increased geographical variation in service provision at the county level. However, when the standard deviation is divided by the mean the resultant value becomes smaller over time, indicating that the counties have become *more* similar in terms of their service levels. That services have become more evenly distributed between 1954 and 1993 is also indicated by the Gini Coefficient which has decreased over time indicating that the distribution of centers in the counties has gradually become closer to the distribution of children.

These measures indicate that over time the network of provision is moving towards equalization in service levels. However, data on licensed capacity across the forty-nine counties in 1993 indicate that there still remain substantial variations among the counties. The least-served county (a relatively rural county in eastern Ontario) had 1.46 child care centers per

1,000 children. The county with the highest service level – 7.86 (five times as many centers per 1,000 children as the least served) is a county where higher education and government services are important components of local employment. Thus, the population tends to be highly educated and employed in professional occupations. In many ways this wide range in service provision reflects the ways that child care services are influenced by class differences and the effects of rural versus urban location.

The 1993 provincial total capacity of 123,374 child care spaces gives an overall allocation of 174 spaces per 1,000 children aged under 5 years old. On a county-by-county basis, the range of service levels varies from about 23 to about 275 places per 1,000, with an unweighed mean of about 130 places per 1,000 children under 5 years old. Figure 5.2 shows the

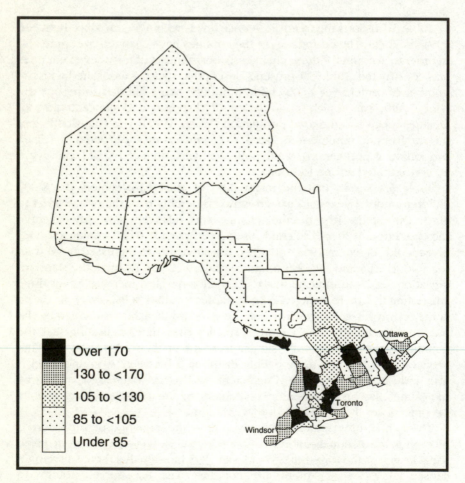

Figure 5.2 Capacity of licensed child care centers, Ontario, 1993

distribution of licensed capacity per 1,000 children under 5 years old in quintiles. While there is considerable variation from county to county, there is also some regularity in service levels. However, it *is* evident that the highest levels are found in the urbanized southern part of the province, particularly in the Toronto area. Only one county in the northern part of the province has a service level in the upper quintile, and in that case only a small number of centers (six) are used by a small population. Elsewhere in the north, and even in some of the more rural counties in the south, the availability of center-based child care is quite limited.

Employing data from the 1991 Census of Canada, further analysis was conducted to examine the association between service levels and three sets of measures: (1) the capacity to implement and manage services; (2) the need for service; and (3) contextual indicators. The capacity to mount services is represented by occupation, education and income variables. These variables were selected as previous research found that leadership in the voluntary sector in Canada is often associated with high socioeconomic status (Armitage 1988; Wharf and Cossom 1987). "Need" is represented by a number of labor force participation rate variables. The contextual variables are population and four regional binary (0/1) indicators contrasting the "Center" (the nucleus of counties around Toronto) with the "Southwest," "East" and "North" regions which are the counties between the "Center" and the extremities of the province.[1] Where data availability permits, gender-specific variables are used.

Table 5.2 reports correlations representing the strength and direction (positive or negative) of the relationship between service levels and each of the variables. For the "capacity" and "need" variables, Pearson correlations are used. For the regional indicators, the mean service level in each region is shown. The coefficients indicate that the variables of socioeconomic status (indicating the capacity to mount services) are generally more strongly correlated with service levels than are the "need" measures. For example, there are high correlations for white-collar employment and post-secondary education and they are highly statistically significant (all but one at better than the .001 level). In contrast, among the "need" variables only the labor force participation rate of women with young children is highly significant (at the .01 level). The correlation for lone-parent families is not significant. This is curious considering that the service represents the main provincial support for child care, and that we might expect service levels to respond to "need" variables more than to socioeconomic status variables. However, these findings suggest that the relationship is the other way around. The contextual variable for population is significant, suggesting an urban–rural difference because the more populous counties are more highly urbanized. There are statistically significant differences in average service levels among the regions. The "Center" and the "Southwest" have relatively higher levels, compared with the "North" and the "East."

Table 5.2 Variable definitions and correlations with service levels

Variable	Correlation coefficients		
	All	*Women*	*Men*
"Capacity" variables			
1 Per cent of white-collar employment	.57***	.47**	.54***
2 Per cent of adults with post-secondary qualifications (gender-specific: per cent of adults with university degree)	.60***	.59***	.56***
3 Per cent of adults without high school	−.54**		
4 Per cent of adults with above average employment income	.30*	.41**	.21
5 Mean employment income		.40**	.29*
6 Mean gross income		.50**	.31*
"Need" variables			
7 Per cent labor force participation	.12	.28*	.08
8 Per cent labor force participation by lone parents	.33*		
9 Per cent labor force participation women with children at home		.17	
10 Per cent labor force participation women with children under 6 at home		.36*	
11 Per cent labor force participation, 1 family member	−.31*		
12 Per cent labor force participation, 2+ family members	.09		
13 Per cent lone-parent families	.24		
"Context" variables			
14 Total population in thousands	.50**		

Note: *** p <.001
    ** p <.01
    * p <.05

Source: Statistics Canada (1992)

Gender-specific data were not available for all the variables and therefore prevents an exhaustive comparison of the relationship between service levels and the "capacity" and "need" variables for men versus women. Table 5.2 indicates that the correlations for women are stronger than those for men. Among the "capacity" variables only per cent of white-collar employment is higher for men than women. Among the "need" variables the only available gender-specific variable was labor force participation and the correlation for men is very low. These results suggest that it is primarily women who are pushing for more child care services.

Table 5.3 Mean number of licensed child care spaces per 1,000
children under age 5, Ontario regions, 1993

Region	Mean
"North"	107.64
"Center"	197.05
"East"	117.00
"Southwest"	140.32
All	129.81
F = 3.21; p < .05	

Note: These are also "context" variables.

Source: Ontario Ministry of Community and Social Services (1993c)

Regression analysis allows for the exploration of the relationship between the child care licensed capacity per 1,000 children aged under 5 (service levels) and the combined effect of all three sets of measures. A number of regression models were constructed to test the joint effects of the measures.[2] The most powerful model accounted for 60 per cent of the variation in service levels. It showed that the most important predictors of service levels are white-collar employment, labor force participation (these two variables are not gender specific data), labor force participation of women with young children, and the binary variables for "East" and "North" (see Appendix II). The model indicated that white-collar employment is extremely important and that as it increases so do service levels. This is consistent with the view that areas where socioeconomic status is high may have a high capacity to organize services. Indeed, of all the socioeconomic status variables used in the correlation analyses, per cent labor force in white-collar jobs was the most powerful predictor in the regression analysis. Turning to the "need" variables, and again rather suprisingly, the model suggested that increased labor force participation by mothers with young children is only modestly associated with increased service levels. The regional measures marking counties in the East and North regions are negatively related, providing further support for the suggestion that these areas have particularly low service levels. Interestingly, the model also showed a strong, negative effect of the labor force participation, suggesting that as labor force participation increases, service provision decreases. One interpretation for this counter-intuitive result might be that where labor force participation is very high, parents use alternative forms of child care. This, in turn, might be because where most adults are working or actively looking for work, they do not have time to volunteer to help implement and manage child care services.

SUMMARY AND OUTLOOK

The availability of child care services is uneven across the counties of Ontario. This chapter has described the rapid growth of services over the post-war period and has described the pattern of inequities that remains despite the expansion of services. The persistence of inequalities would appear to be a pervasive result of a decentralized delivery system in which the development of service depends on the success of initiatives in localities. The regression results suggest a very strong influence of occupational status on the availability of formal child care service: where there is a high proportion of the labor force in white-collar occupations, service levels are also high. This can be understood in reference to the administrative and bureaucratic skills required to organize and implement services in a decentralized system, although it also raises concerns about the equity of distributing public resources in this way. The finding that services are strongly and negatively related to labor force participation may suggest another limitation with decentralization: that it may not be practicable for parents to take the time to be involved with the volunteer work needed to mount services. The strong negative results for the regional indicators suggest that people in these traditionally under-served areas continue to suffer low access. A fruitful question for further research would be to examine the other forms of child care used in the under-served areas. It would also be useful to assess whether their limited development of child care services in these areas is paralleled in other parts of the welfare state, for example, assisted housing; and if so, how people engage the services they need.

Policies addressing child care have concentrated on controlling and regulating providers and on putting subsidy mechanisms in place; they have, for the most part, not addressed spatial inequalities. Under recent policy developments the provincial government intends to identify targets for expansion in specific areas and to provide developmental resources; but it remains to be seen whether these efforts will be able to direct expansion geographically.

Part IV

THE LOCAL LEVEL: JOURNEY TO CHILD CARE AND COPING STRATEGIES

6

THE JOURNEY TO CHILD CARE IN A RURAL AMERICAN SETTING

Holly J. Myers-Jones and Susan R. Brooker-Gross

INTRODUCTION

The growth in the labor force participation of married, middle-class mothers has made child care a necessity in the US. Today, an unprecedented number of women, with children under the age of 5, are working outside the home, but so are the grandmothers, aunts, and cousins of these children, making informal, extended family-based child care arrangements less available. Add to this the increased geographic mobility of the labor force, and we begin to understand the necessity for the creation and maintenance of formal child care arrangements as fewer working mothers can rely on pre-existing, neighborhood-based child care.

While political and social rhetoric has expanded child care from a "women's issue" to a "family issue," daily lives of many women and not quite so many men are bound up with taking and picking up children to and from their sitters, child care centers and schools, usually as an add-on to the work-trip. Thus work-trips are increasingly multi-purpose trips, and for many mothers, part of the "second shift" of domestic work (Hochschild with Machung, 1989).

From the standpoint of transportation geography, multi-purpose trips add complexity to our geographic models and present new challenges to the transportation planner. But the journey to child care is more than a simple add-on to the journey to work. The journey to child care is dependent upon the geographic configuration of job, child care and residential opportunities in the local area. This chapter examines child care trips in a rural community in southwestern Virginia where job and child care opportunities are limited. Our motive in studying these trips was twofold: (1) to examine the ways individuals incorporate the additional complexity of non-home-based child care into their daily routines; and (2) to examine how child care trips differentially impact different household members and social groups.

77

CHILD CARE OPPORTUNITIES

Child care varies widely in type, level and quality. Some providers offer little more than child minding in the most conventional sense. Others, often described as preschools, primarily have an educational mission, while still other providers serve children with special needs. While there has been persistent concern in the US that paid child care does not adequately promote the development of very young children, some centers are models of quality care (see Cromley, this volume). Formal child care centers may be for-profit or non-profit. Some programs are subsidized by employers or governmental programs. Others require cooperative support from parents, a form of in-kind subsidy. In addition to formal child care centers, parents' options include hiring individuals to come to their home, or taking children to in-home child care providers, or asking extended family or friends to look after the children while parents work (see Bloom and Steen, this volume). Private preschools and after-school activities add to the diversity of child care. Lack of access to child care results in unattended, "latch-key" children or maternal unemployment. With different locational patterns of services, a complex geography of opportunity and access results.

TRAVEL ISSUES

The proximity of child care locations to workplaces needs to be considered. If children are near home but far from parents during workdays, emergency trips will be difficult, as will spontaneous visits to see the children or to check their typical day's experience and the caregiver's performance. In many instances, the operating hours of centers also lead parents to choose from options close to work rather than close to home. On the other hand, proximity to the parents' workplace may mean a long journey for the child, exposing her or him to the dangers of commuting and its tedium. As children grow, the need for child care typically changes from full-time to before and after school during the school year, back to full-time during the summer months. When children begin school, parents may experience a reconfigured set of child care options which has to accommodate school location as well. The journey to child care must include consideration of the work-trips of parents and the relationship of these trips to the children's journeys.

The journey to child care must also accommodate the capability constraints of children at different stages of their lives. Sleeping hours and eating times are less flexible for most children than for most adults, and not flexible at all for infants. Even "potty" stops are inflexible for younger preschoolers. How long can an infant or young child ride in a car or bus? How early can a child be awakened in the morning? Can a child safely nap on the way home from child care while buckled into a safety seat? And parents worry about the psychological impact of a lengthy, daily separation from parents.

The journey to child care complicates the journey to work. Hanson and Pratt (1988) argue that the two are not separate "strings" connecting two or three nodes, but part of a continuous spatial context. The work-trip encompasses the context of the workday routines of parents, and the interplay of spatial choices within particular local structures: home, work, child care and the neighborhood settings. Studying spatial choice behavior becomes problematic when few opportunities are available in any one of these aspects of daily life, and opportunities may be especially limited in the journey to child care (Tivers, 1985, 1988). We assume that both home and job locations are more fixed than is child care site choice, although the influence of child care on either the home location decisions or job choices should not be overlooked, especially if employment benefits include either on-site care or financial aid for child care. Hanson and Pratt (1992) suggest that for many women, job location may be constrained by child care options, rather than the reverse. Choices for child care include many different kinds, but are often limited in number. Choices are limited by child care cost and scarcity, and variations in perceived quality. In areas where child care demand exceeds supply, the consequences include latch-key children and maternal under- and unemployment. Under-employment often takes the form of part-time work, rather than full-time. Part-time employment may allow a mother of school-aged children to be home in the afternoon, or may be a part of a household strategy to construct sequential work hours for husband and wife (Pratt and Hanson, 1991b; Fine, 1992; Rose, 1990).

The hours of child care provision are also critical. Child care centers must be open longer than the typical work shift in a locality, with hours planned around the timing of that typical shift. In Palm and Pred's (1978) hypothetical case of a mother with two job opportunities and one child care option, the mother's job options were narrowed to one when the coupling constraint of child care hours was considered (also see Pickup, 1984). The more distant in time–space a child care center is from places of employment, the earlier it would need to open and the later it would need to close. These coupling constraints may limit choice even in environments seemingly rich in child care opportunities, and make it difficult for parents with non-traditional work hours (for example: swing shift, evening and weekend hours, rotating shifts) to meet their child care needs. For example, in an immigrant community in Montréal, Rose and Chicoine (1991) found that women's limited access to jobs resulted in work hours that made use of formal child care difficult.

Given the complexities involved in making child care choices and in building the child care link onto the journey to work, should we expect to find any spatial regularities in the journey to child care? We can speculate based on findings about the regularities in the journey to work. Both the space–access trade-off models and job search models have ample empirical evidence of higher-income workers traveling longer distances to work than lower-income workers. Similarly, women workers typically travel shorter distances than do male workers, because women too are "disadvantaged" workers, whether through lower household incomes, or through gendered definitions of household responsibilities, including "mother" and "wife," which require closer proximity to home or extra hours at home. Women are also more likely to be disadvantaged due to a lack of transportation alternatives (Johnston-Anumonwo, 1992; McLafferty and Preston 1991; Pickup, 1989; Preston *et al.*, 1993; Rosenbloom, 1988).

Some anomalous cases arise when opportunities for work and/or housing are restricted. African-Americans have typically traveled farther to work than occupationally similar White Americans (Deskins, 1973; McLafferty and Preston, 1991; Preston *et al.*, 1993), particularly for African-American women compared to white women (McLafferty and Preston, 1992); some male part-time workers may travel farther than full-time workers (Hanson and Johnston, 1985); and women do travel long distances when nearby jobs are not available (Brooker-Gross and Maraffa, 1985; Rutherford and Wekerle, 1988; Kipnis and Mansfeld, 1986; Pickup, 1989). The anomalies arise both from restrictions on particular groups in society and from particular limiting geographic contexts. The probable alternative to longer-distance commuting in each of these cases is less than full employment.

A model of journey to child care that fits the space–access trade-off concept would predict short, distance-sensitive trips to take children to a center or childminder, with shorter distances for economically or domestically disadvantaged workers. However, if the opportunities for child care are highly constrained, a longer journey may be necessary. The alternative may be no employment for one parent who then cares for children at home. Macro-level studies of the availability of formally organized child care have indicated that women's labor force participation is influenced by the abundance or scarcity of child care options. In comparison with many other northern European countries, the UK has fewer child care options, and lower-paid labor force participation of mothers of young children (Bowlby, 1990; Fine, 1992). Modeling of child care costs in the US indicates parallel results. Higher costs of market-based child care have a negative effect on married women's labor force participation, and given that a woman works, high child care costs lead to higher probability of using informal child care rather than market-based care (Blau and Robins, 1988). Ironically, the reliance on lower cost, informal child care by lower-paid workers makes it difficult for many such workers to take advantage of child care tax breaks available in both the US and Canada (Bloom and Steen, this volume; Truelove, this volume; Rose, 1990).

Just as the provision of child care is a factor in women's labor force participation, so too is the geographical accessibility of that child care. A 1975 study by Darian found that "convenience" of child care increased the labor force participation of mothers with young children, indicating that short trips to child care are prerequisite to many mothers' employment. Similar results were found by Stolzenberg and Waite (1984). More recently, Pratt and Hanson found that women in their sample in Worcester, Massachusetts, evaluated job opportunities with the proximity of schools and child care facilities in mind (Pratt and Hanson, 1993).

The relationship between socioeconomic status and length of child care trips in the few studies done to date has been ambiguous. In one California study (Women and Environments, 1988), the distance to child care increased as socioeconomic status increased, while a University of Connecticut survey (Cromley, 1987) showed a preponderance of short-distance trips by high socioeconomic status users. The Connecticut study showed that higher-income, university faculty lived near the job site, and used the relatively expensive child care center. In Michelson's (1983) study in Toronto, single parents were more likely to use child care centers closer to home than married parents, although the time of child care trips was greatest for single parents who used public transportation.

Michelson (1983) found that child care trips added significant distance to the total work-trip for some parents, mostly mothers. At least 28 per cent of child care trips were not on the direct route between work and home. The extra distance was greatest for users of formal child care centers and smaller for users of in-home care. Formal centers' larger scale puts them in neither residential nor workplace locations, while neighborhood caregivers were often selected because of their proximity. Trips to grandparents as child caregivers added an intermediate distance to work-trips because, as Michelson said, "they are where they are" (1983: 96). Relatives too, however, need to be close enough, as D. Rose (1993) has noted. Michelson further showed that total mileage was less significant than added mileage in explaining reported tension by parents.

These seemingly contradictory results concerning both the journey to work and the journey to child care indicate the importance of the local context in understanding both. Although women's participation in the labor force is increasingly an economic necessity, such participation is highly contingent upon the social and economic construction of their residential, work and child care communities (Hanson and Pratt, 1992; Rose, 1990). These communities vary greatly in size, level of urbanization, class dimensions, racial and ethnic composition, and availability of employment opportunities. To understand how working parents cope with the complexities of their daily lives, one must understand the local context which both presents working parents with various opportunities and serves as the material from which parents can construct and reconstruct new coping strategies and opportunities as the need arises.

THE STUDY AREA AND SAMPLE

The town of Blacksburg, Virginia, is the geographic context for this study. On a small plateau at the base of the Appalachian ridge and valley region, it is the site of Virginia Polytechnic Institute and State University – Virginia Tech. The university is the primary economic base of the town, and while urban development in and surrounding the town has created suburban residential subdivisions and campus-like factories, the region is embedded in the high poverty area of the Appalachian region. The university itself expanded rapidly in the 1960s and 1970s to a current enrollment of over 20,000 full-time students, a population figure included in the town's 35,000 population. Because of the recency of this growth, the town's residential geography appears at first glance to be topsy-turvy. Older apartments built for veterans returning from the Second World War occupy inner zones, along with still-gracious neighborhoods of single family homes that pre-date the university's expansion, and similar neighborhoods of the early expansion years. The latter are sometimes known as "faculty ghettos." Large apartment complexes built to accommodate recent enrollment growth, and newer subdivisions of single family homes ring this inner zone.

Beyond the town limits, two kinds of residences are home to other university employees. In neighboring small towns and their suburban subdivisions, junior (and humanities) faculty and mid-level staff reside in housing outside the intense price competition and inflation of the node of Blacksburg. Senior faculty (and business and engineering faculty) live alongside the doctors and lawyers in low density, mountainside exurbs, some very close in distance to the town itself. A final handful of Blacksburg workers has adopted a more rural lifestyle on farms and hobby farms in the countryside.

The employers who complete the picture of work in Blacksburg include retailers who employ residents full-time, but who also rely on substantial student labor, and the schools and hospital which have a notable dependence on tied-migrants – especially wives of male faculty members. A few industries in high-tech manufacturing draw from diverse residents of the community. Finally, Blacksburg also acts as a bedroom community for nearby rural-based industries. The US Army contracts a site for manufacturing explosives several miles west of Blacksburg, and at the time of the study, a sizeable facility of AT&T was operating between Blacksburg and the college town of Radford, Virginia, 12 miles away.

While small in size, Blacksburg offers a complex set of relationships among occupational class, household class, and age and gender. The university and the industries bring a large, well-educated, middle-class population to a piece of Appalachia, a region known for its dependence on extractive industries, low wages and lower levels of education. However, census statistics on poverty highlight neighborhoods of students – undergraduates with little to no income, but living middle- to upper-middle-class lifestyles. Underemployment is characteristic of middle-class married women, who take clerical jobs for which they are overqualified as their only option in the town's limited job niches. These women compete for jobs with, and work side-by-side with others who are longer-term residents of the region, many of whom are the primary earners in their households.

The diversity of child care mirrors the class diversity of the population. For this study, we examined one slice of the diversity in child care: formal child care centers. Included are both for-profit and non-profit centers that serve a non-specific clientele; that is, they are not tied to a specific employer or program. Even among such centers, services vary along many dimensions, including costs, licensing requirements, staff preparation, activities provided, physical quality of buildings, playgrounds and the ages of the children cared for. Spatial equity questions arise as service provision is compared with proximity to residential neighborhoods, themselves differentiated by family composition, economic status, race and ethnicity.

The three centers examined for this study were the largest in the town, each serving approximately 100 children. As we shall see, upper middle-class families predominate. Children from lower-income families are effectively priced out. Subsidies for child care come largely from the State of Virginia or from local churches, but few children in these three centers received such subsidies, although the centers did offer multiple child discounted prices.

While centers in the region and beyond vary widely, the three examined here were more similar than dissimilar: two centers were for-profit, and one was church-owned and operated. Two centers offered before- and after-school transportation for elementary school children, and the third ('center') was directly adjacent to an elementary school. All offered enrollment to children from 6 weeks old to 12 years old. The location of the three centers highlights Blacksburg's string-town morphology, with the oldest facility ('center') centrally located near the university and near the town's commercial core, the next oldest on the north end of the main street ('north'), and the newest just off the southern arterial ('south'), near the university's new research park and closest to the town of Christiansburg, a popular alternative residential location for many of Blacksburg's workers.

The results of the study come from a sample survey that was distributed in person during drop-off and pick-up hours at the three centers, and returned to boxes at the centers. Parents or others who were responsible for the children's transportation were given the forms. While school buses transported before- or after-school children on one leg of the trip, no center allowed children to leave unaccompanied. The survey was administered in the spring of 1988. Respondents filled out questions about their own responsibilities for child care trips, as well as parallel information on any second person with routine responsibility for child care trips. Respondents had an opportunity to comment on the survey form. Some of these comments are cited below. A total of 129 surveys were returned for an average return rate of 50 per cent for the three centers combined.

FINDINGS

The trips made to child care were short, with over three-quarters of the respondents living and working in Blacksburg, including the 57.5 per cent of respondents who worked at the university. Mothers were most often involved in the child care trips. Rather than ask for percentage of trips by each parent or other caregiver, we asked who had routine responsibility for dropping off and picking up the children. Even with this open definition, mothers were reported as solely responsible for dropping off the children by 47.7 per cent of respondents, and for picking up children by 54.6 per cent. Both parents were reported as involved in dropping off children by a smaller number of respondents (23.8 per cent), and picking up children (30.8 per cent). Compare our results with a 1993 Los Angeles survey that showed mothers' child care stops during work-trip commuting to be three times those of fathers' (Nichols, 1994). Clearly, Blacksburg is quite different from Los Angeles, but this shows that gender-based activities transcend locality.

Respondents were asked why they chose the center they used, and results predictably ranked quality of care first (85.4 per cent). This variable is of utmost importance to parents and ranking varied little among the centers. In this college-based community, the level of care in the centers was indeed high, and child development professionals at the university occasionally engaged in consulting with centers, while students in the early childhood development degree program worked as interns. Specifics of programs varied among the three centers, as did the quality of facilities – one in very old quarters and one in new, purpose-built facilities. The absolute variation of the three was relatively small, however, as the quality of center response demonstrated.

The next factor chosen most often was the educational value of the center (69.2 per cent). Again there was little variation among the centers. Cost mattered to 34.6 per cent of the survey respondents. Recommendations from friends were cited by 46.9 per cent, and recommendations of professionals by 18.5 per cent, with the oldest center respondents citing friends' recommendations the most, and the newest center with the most recommendation from professionals. "Space available" was expected to be important, since all three centers had active waiting lists. It was not among the highest rated factors, averaging 33.8 per cent. This factor may have been taken for granted by the respondents. In a similar vein, dissatisfaction with previous child care arrangements was rated as important by only 25.4 per cent of the respondents.

Given that these three centers comprised the largest supply of child care available in Blacksburg at the time, that the city itself is small, and that child care availability is insufficient to meet demand, we expected the sensitivity to distance to be of limited significance. Child care choices are highly constrained, and differences in distances to the three centers are small. Yet over one-third of the respondents (35.4 per cent) checked distance from home as an important factor, and clients of the "center" facility cited distance from home more often (46.2 per cent) than did clients of the other two locations. When asked about distance from work, respondents at the "center" location again rated distance more (33.3 per cent), while the "north" child care site, farthest from campus, had the lowest rating on distance from work (20.4 per cent). Forty-one per cent of the central facility's respondents also checked distance from home as an important factor, ten points higher than either of the other centers' clients.

Table 6.1 (part A) shows the mean straight-line distances from the respondents' homes to their child care center. The shortest distances are lowest for each center's own respondents (compare within columns), although the "center" child care site was closest to all groups (compare within rows). The locational advantages of the "center" facility figure in distance from work, too (Table 6.1, part B). The "center" site is the closest to

Table 6.1 Aspects of the journey to child care in Blacksburg

A Miles from home to child care centers

	To center		
	North	Central	South
North center users	1.85	1.75	3.01
Central center users	2.09	1.40	2.83
South center users	2.66	1.96	2.47

B. Miles from work to child care centers

	To center		
	North	Central	South
North center users	2.02	1.04	2.30
Central center users	2.14	1.26	2.60
South center users	2.49	1.47	1.89

C. Miles added to work-trip by child care trip

North center users	1.88
Central center users	1.41
South center users	2.47

all the respondents' workplaces, but the two non-central facilities are nearer the workplaces of their own users than to each others' clients.

A final distance factor is the mileage added by the journey to child care (Table 6.1, part C). In Michelson's (1983) study, the tension of child care trips was more strongly related to added mileage rather than the absolute distance involved. The advantage of the "center" facility in this respect is clear, followed by the "north" center, and the "south" center users face the greatest miles added to their work-trips. Overall, the maximum added mileage was 6 miles and the median added mileage was under 2 miles; 4–6 miles would compose a cross-town trip. That so few are involved in such back and forth trips is testimony to the sensitivity to distance in the journey to child care even in a town small enough to intuitively be a frictionless zone. Respondents' comments revealed the tension potentially involved in the trips to child care: one said that the picking-up and dropping-off trips always took longer than planned, and another commented that the morning trip was made difficult because of the need to be at work, and the afternoon trip by the need to get home for dinner. The time for transition for the children to shift from one setting to another adds a dimension to child care trips that is absent from tasks such as picking up dry-cleaning or groceries. The teary-eyed child who resolutely clings to a parent in the morning may be the same one who cannot understand the unreasonable parent who pulls him or her away from friends and play in the afternoon.

A respondent's comment that commuting to child care is a good setting for one-on-one discussions with the child reflects a positive view of the potential quality of the child care trip. Another respondent found different side-effects of child care. She commented that running household errands before picking up her child had deprived the youngster of worldly experiences: her son had no knowledge of drive-through bank windows until he was too ill to be at the center and accompanied her on errands.

OCCUPATIONAL STATUS

The primary surrogate for socioeconomic status in this study, and indeed, in Blacksburg, is occupation. Since the university is the dominant local employer, professional occupations are abundant. Not surprisingly, among the child care users surveyed, professional occupations were the largest category (52.3 per cent). Only secretarial (10.8 per cent) and administrative jobs (14.6 per cent) accounted for more than a handful of respondents. Students were another small proportion of child care respondents (13.8 per cent). Of course, the users of the three centers are not representative of potential child care users in the community: professional occupations are over-represented. Without the added mileage of child care trips, average work-trips in this setting are short for professionals. Generally, higher-income households disproportionately use the formal child care centers, making short trips in the process, with proximity to both work and home.

Some university employees, as well as employees of other local employers do live outside Blacksburg and travel long distances to work. For the university's workers, we know that proportionately more of the longer distance commuters are non-professional workers (Maraffa and Brooker-Gross, 1984). However, out-of-town users of the centers are predominantly professional workers. Among the child care survey respondents, 50 per cent of the Blacksburg residents are professional workers, while 60.7 per cent of the non-Blacksburg residents are professional workers. In short, out-of-town non-professional residents who work in Blacksburg are under-represented in this survey of child care center users.

The preponderance of professional workers among the child care respondents makes it difficult to break out specific occupational categories. We can, however, compare professional families to the remaining households. For this analysis, professional families include those families where either the respondent, or the other person identified as responsible for child care transportation, listed a professional occupation. In the total sample, such families comprised 63.1 per cent, with little variation among the centers (from 61.5 per cent to 65.3 per cent). The advantages accruing to these families include short trips to child care, with an average of 1.8 miles from home to child care, compared to the slightly higher 2.0 miles for non-professional families. However, the non-professional occupation families had shorter trips from the child care centers to their place of employment than did the professional occupation families: 1.4 miles compared to 1.9 miles.

DISCUSSION

The study has shown that mothers in the sample have primary responsibility for child care trips, even with a very open definition of responsibility, and despite the above average occupational status of respondents. Mothers' work-trips are more closely linked to child care than are fathers'. Parents are more likely to share responsibility for picking up children than for dropping them off. This finding is consistent with Rosenbloom's study (Rosenbloom, 1993) of linked trips in Austin, Texas. She found that married mothers consistently had a higher proportion of linked trips on the way to work than did married fathers. In contrast, fathers had a consistently high level of linked trips after work, with mothers' linked trips varying with the age of youngest child and with marital status. While Rosenbloom's data do not tell us which of these trips are to take children to babysitters or child care centers, the patterns are consistent with our findings.

It is possible that the household's coping strategy depends on the mother's work hours and/or flexibility in morning work hours. More women than men in our survey have "regular" work schedules (76.4 per cent compared to 68.4 per cent). Of those with regular daily schedules, 87.5 per cent of the women and 94.2 per cent of the men work 7.5 or more hours per day, indicating that women are engaged in more part-time work than men. Mothers' more regular work hours may bring stability to the child care trips. For the part-time workers, the added available time may ease the stress of child care trips. On the other hand, fathers' apparent flexibility in making the morning child care trips seems not to influence which parent is responsible for taking the child or children.

The different patterns between morning and afternoon sharing of child care trips may point to a greater stress level for women's share of the trips. Arriving at work at a scheduled time is a more precise target than arriving at home in the evening, and children are more likely to have stressful reactions to being left at a child care center than to being picked up. A potential major stress-point in the evening is the closing time of the center. All centers had clear and strict financial penalties for arriving after closing time, not to mention the stress of a child who has not been picked up when her or his peers have all left the center.

There were, of course, large numbers of fathers who were involved in the child care trips, a factor that may also reflect the occupation and class structure of the community. Anecdotally, fathers and mothers were part of complex strategies of family commuting. Two respondents noted that in a family of two parents and two children, each parent was responsible for one child's journey to child care. One commented that each parent had responsibility to prepare that single child for his or her day, minimizing the intra-household rush. The other respondent remarked on split duties, noted that each child had a different destination, one toddler to full-time child care, one school-aged youngster needing before-school care. It was a practical matter for each parent to deliver and pick up one child. Blended families had even more complicated strategies. For example, one respondent commented:

> These children are brought to daycare from [a town 10 miles to the north] by their stepfather. The youngest child stays during the morning and is routinely picked up by his stepmother or father. The older child usually stays only before school and rides the bus home to his father's house. Both children are picked up by their mother after work and return to [the home town].

There was the intriguing handful of still other caregivers. One person who completed our questionnaire enquired about the "Who has routine responsibility for dropping off and picking up the children" nature of our questions' phrasing. She was a substitute – not the regular person who picked up the children, but it was her job as babysitter when the parents were out of town. Other parties responsible for chauffeuring the children were private babysitters and grandparents. Each center was careful about who was allowed to pick up children, so that exceptions to the routine had to be negotiated with the center and authorized in writing.

A second finding is that distance minimization is an important factor. While the empirical finding of this principle of least effort is expected, and while we need to remember that some households engage in cross-town, out-of-the-way trips to centers, it is surprising that distance sensitivity appears in a small city with few choices of either child care or of employment. We had undertaken the study fearing that the results could be geographically uninteresting. In addition to the small scale of absolute distances, there is little reason to expect marked neighborhood effects in choice of center. With the supply of center care insufficient to demand, real choices are highly constrained. The emergence of distance sensitivity is consistent with anecdotal cases of newcomers to town choosing a residence based on proximity to child care centers. We may also be seeing the effects of the newest center ('south') located at the rapid growth periphery of the town, and the oldest center most reliant on word-of-mouth recommendations from friends and neighbors. Our findings reinforce those of Darian (1975) and Stolzenberg and Waite (1984) that convenience of access to child care is important in mothers' labor force participation.

As expected, we have found that the users of formal child care centers are above average in occupational status. Center prices tend to be intermediate between child care providers who come to the client's home, and providers who mind children in their own homes. While the structure of the centers may appeal to professional workers, it is also true that the costs of formal centers fit the budgets of more highly paid workers. Few provisions were made in the centers for lower-income families, the church-run program being the only one of the three offering subsidized programs. While transportation was a trivial issue to some of the respondents, one comment underscored the need for accessible child care, accessibility in both price and location:

> My concern is regarding next year. My child will be starting kindergarten actually, and as you can see, I'm at [the university] all day, either in class or working. This daycare [central location] is one of two that my child, as an AFDC recipient, is eligible to attend. No transport is provided from the school to the daycare after school. As you can imagine, this is going to cause me considerable problems. Besides I am confident that I am not alone in this predicament.

There is an elementary school directly adjacent to the "center" facility mentioned here, but the school assignment for this child was to a different school!

We should also recognize that the professional labor force is usually recruited from a national labor pool. Parents native to the area have a greater potential of having their own relatives near enough to provide care, and better access to informal networks of service provision. This situation is analogous to Hanson and Pratt's (1992) findings that locally oriented women in Worcester, Massachusetts, found their jobs through informal local networks. By contrast, those recruited to the Blacksburg region, whether managerial or professional workers of the university and the major manufacturing facilities, or students (especially graduate students), are unlikely to have kin close by, and may rate center-based care more highly or be unaware of other options. The number of long-term residents who are available to become child caregivers may also be proportionately smaller in the university town, with a high rate of women's labor force participation coupled with college enrollment.

Although lower-income workers' children were under-represented by this study, their very absence leads to some hypotheses about the interrelationship between housing, employment, and child care choices and locations. As the largest employer in the locality, the university has responded to unmet child care needs for its staff, faculty and students, by establishing a child care referral center in 1989. (The university does not have a full-time, on-site child care center, although the child development program runs a part-time center.) The introduction of a referral center at the university reflects the national trend of increasing numbers of this service (see Bloom and Steen, this volume). A key activity in the referral center is the fostering of new in-home providers. This response reflects the bi-polarity of class in the region: in-home care is a badly needed service and encouraging its expansion offers economic opportunities for potential providers. The referral list of providers typically lists more providers in the county seat town of Christiansburg, 10 miles distant from Blacksburg, than in Blacksburg, despite the smaller population size of Christiansburg (personal communication, Theresa Reed, Referral Center staff). Some university employees and students live in Blacksburg and take children to Christiansburg providers, while many others are Christiansburg residents. Furthermore, the opportunity to become an in-home provider may be more attractive in a town less dramatically influenced in its demographic and occupational structure than a college town. The price of housing in Christiansburg is less driven by the competitive market in Blacksburg, making it a realistic residential choice for lower-paid workers, and a more realistic choice for business opportunities as in-home providers of child care.

The supply side of child care – especially in-home care – can be highly responsive to local economic changes. In Pratt and Hanson's (1991b) discussion of sequential care by parents, their study area had a booming economy at the time, with shortages of workers to fill female-dominated jobs in clerical, hospital and other service work. While they do not directly address the availability of in-home caregivers, the description suggests that more lucrative jobs were available to women. Similarly, the availability of caregivers varied across the relatively small space of our locality, reflecting the uneven landscape of women's labor force participation and their university attendance rates (both of which are greater for Blacksburg relative to the region).

This situation brings more complexity to expectations of the distance sensitivity of child care trips. If the upper middle-income families among the respondents were surprisingly sensitive to distance, would lower-income families be even more sensitive, as national work-trip data suggest? Alternatively, the cost structure of child care may overwhelm costs attributed to distance, and result in longer child care trips for lower-income families. England (1993a) has suggested that the importance of a single parent's income may lead to the necessity of longer trips to work. By extension, the longer trips may include longer child care journeys. England further makes the point that a complex web of sociospatial relationships characterizes women's journeys to work, including "good child care arrangements, the local school system, and low mortgage payments" (1993a: 237).

A year after this study, the complex web of this locality unraveled suddenly, with the closing of the oldest, centrally located center at short notice. A similar crisis had occurred three years earlier when the only employer-sponsored center in the area closed. The options put together by users of those centers highlighted not only the *difficulties* facing parents needing child care, but also the *creativity* with which parents resolved these problems. Reportedly, some children enrolled in other centers or after-school programs; some private sitters were found; and some mothers quit work. Multiple options were part of the reported strategies. Such a crisis serves to exaggerate the normal difficulties families have in finding child care. These types of problems may be even more severe in non-metropolitan settings such as ours, where both child care and employment options are spatially limited.

The current situation differs from the time of the study. One of the centers we studied has since relocated to a large residential subdivision, improving its physical facilities in the process. Another center opened in a purpose-built building, located in a shopping district, and a near-town center dramatically changed the nature of its service from mostly secular to strongly religious. The university's referral service continues, having been established, in part, as a compromise effort to balance the demands of faculty and staff for an on-campus center and the economic interests of local child care providers. All of this indicates that the stability of child care provision implied by our snapshot survey is itself an illusion.

This study in a non-metropolitan environment raises the issue of geographic scale. We were, as we indicated earlier, surprised at the geographic sensitivity generally found in our survey despite a highly constrained situation. Of course, we may be seeing an artefact of small city life: low expectations of daily travel distances. One respondent commented on the travel expectations in this setting. Although he used the "north" center, he lived in the southern end of Blacksburg, worked in Christiansburg to the south, and his wife worked in the western end of Blacksburg. He commented:

> It seems totally out of the way for us, as we have no other business in extreme north Blacksburg. In reality its only a couple of miles from one end of Blacksburg to the other, and the actual inconvenience is minimal compared to larger towns or cities.

A larger urban center would have more child care options, more dispersed residential and employment patterns, and potentially a greater acceptance of the necessity for longer daily trips. Larger cities also have a greater degree of land-use and socioeconomic differentiation, and pose greater difficulties in sorting out the interrelationships between spatial structure and household choices.

In summary, the survey reported here confirms general trends found in studies elsewhere: that mothers' responsibility for child care trips is greater than fathers' responsibilities, and that gendered definitions of responsibilities are a more likely cause than are the logistics of daily employment schedules. We also found similarities in that respondents valued short journeys to child care. Perhaps more importantly, this study confirms that the addition of child care trips for many families requires the development of elaborate coping strategies which structure daily life, and that individual family choices cannot be predicted on the sole basis of child care availability either in an economic or a geographic sense.

Differences in our study from others underscore the importance of the local context, both geographically and in the occupation and class structure of the community. Class differences are expressed both in the ability to pay for specific types of child care options, in this case center-based care, and in access to other opportunities. Access to non-center alternatives is related to one's status as native or in-migrant, as well as ability to pay. Rural, small-town, and non-metropolitan areas typically have lower rates of women's labor force participation than do cities, a factor that affects both the demand and the supply of child care options. They also have more limited job opportunities, and those that do exist are likely to be more locationally limited. All of these factors affect the complexity of the daily routine needed to accommodate the child care trip, and in turn, impact female labor force participation.

7

THE LOCATIONAL CONTEXT OF CHILD CARE CENTERS IN METROPOLITAN TORONTO

Marie Truelove

INTRODUCTION

The provision of child care services in cities presents an opportunity to examine the spatial dimensions of a social service that is provided both publicly and privately. Geographical access to a service must be regarded as an important component of the effectiveness of that service. The problem of unequal access to a social service such as child care is important for both policy makers and researchers – and, obviously, for the consumers involved. There are strong links between the locational and political issues involved in child care. Geographical analysis can contribute to an understanding of the effects of present and proposed policies on child care, particularly their spatial and redistributive impacts. For instance, Canada's mixture of ownership types – municipal, non-profit and commercial – has interesting implications for service provision and access, especially as non-profit and commercial centers often have quite different spatial distributions within cities. Thus, the actions of government and non-government agencies are an important set of constraints under which parents' choice of child care center is made, even when such actions are intended to broaden choice. Access to child care includes the institutional factors such as regulations and policies that make up the criteria for eligibility for admission to the child care system, the level of government funding, the prices charged, and waiting lists (priorities) for child care spaces (see Truelove, Chapter 3, this volume). Government agencies have had major impacts on the availability of child care, and especially on the availability of subsidies (one of the greatest difficulties for many families is that few new subsidized spaces are being created). Changes in policies can create major changes in the supply and demand for child care. Thus the diffusion of child care centers is dependent on the agencies and providers and their creation of the service. The role of institutions in the provision of child care in Metropolitan Toronto is one focus of this chapter.

Clearly, families do not have an unconstrained choice of child care centers, and the center chosen is not only affected by institutional factors but other issues as well, only some of which can be investigated in this chapter. At one level, the time-geography of the situation – comparing parents' workplace location and working hours with the location and operating hours of child care centers – may also severely affect access to this service (Martensson, 1977). For example, Michelson (1985) found that child care centers are often difficult for parents to reach and are more distant from home and work than other, less formal kinds of child care such as assistance from neighbors, relatives and other sitters. More broadly, government and non-government agencies have some control over where, and for whom child care is available; this, in turn, impacts on parents' use of center care versus other types of child care, as well as on the distances children travel and the prices parents pay.

In short, institutional actions can help to explain both the spatial patterns in the supply of centers and individual behavior patterns (Brown, 1981). In this chapter both these patterns will be explored using Metropolitan Toronto as a case study. First, the spatial evolution of child care since the early 1970s will be described. Second, in order to understand some of the ways that institutional actions influence the daily lives of parents, various aspects of travel patterns to child care centers will be examined.

THE SPATIAL EVOLUTION OF CHILD CARE CENTERS IN METROPOLITAN TORONTO

Metropolitan Toronto consists of six local municipalities: the City of Toronto and the two small municipalities of York and East York are often described as the "inner city"; while the cities of Etobicoke, North York and Scarborough represent post-war suburbs. Their combined population is 2.3 million and Metropolitan Toronto covers an area of 241 square miles (see Figure 7.1). Over the last twenty-five years formal child care has grown dramatically in Metropolitan Toronto. The 1970s were an important period in terms of the expansion of child care, and one contributing institutional factor was the great expansion of subsidies during that time period. In 1968, there were only 39 child care centers; in 1971 there were 104 centers; in 1981 there were 266; and by 1992 there were 497 centers providing year-round, full-time care for approximately 25,500 children (of these spaces in full-time centers primarily aimed at preschool children, approximately 3,500 spaces were for school-aged children) (Community Information Centre of Metropolitan Toronto, various years). Despite this, a conservative current estimate for child care need for children under 6 years old is that 50,000 children may be competing for the 22,000 full-time child care center spaces in Metropolitan Toronto.[1]

To examine the evolution of child care centers in Metropolitan Toronto since 1971 data were assembled from an annual publication of the Community Information Centre of Metropolitan Toronto, which began

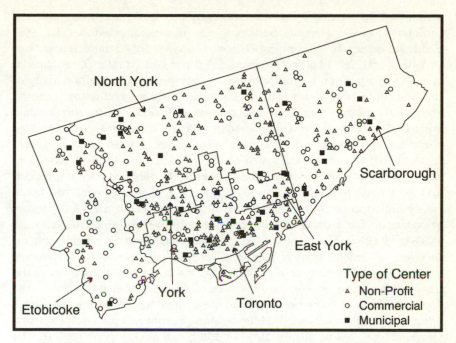

Figure 7.1 Child care centers in Metropolitan Toronto, 1992

publication in 1969/70. This annual directory, *Day Care*, provides the location, proprietary type, capacity,[2] ages of children accepted, and the existence of purchase-of-service agreements (approval to take children with subsidies, many of whom live in lone parent families) for all licensed child care centers in Metropolitan Toronto. Although centers may have opened, closed, moved or switched from commercial to non-profit (or vice versa) any time after the data are collected, the degree of error is small. The Community Information Centre obtains the information from the province and checks it. *Day Care* has not been published each year recently, but an up-to-date computer listing of centers was obtained from the Community Information Centre in February, 1992. Tables 7.1 to 7.3 show only those centers providing full-time care (i.e. for children under 6 years old). Of course, not *all* spaces in any particular center need be full-time, but the centers that serve school-age children *only* have been excluded.

An unusual attribute of child care centers in Canada is the mix of ownership types: non-profit municipal centers (of which only Ontario has any significant number); private non-profit centers, including parent-run cooperatives; and for-profit commercial centers. This mix, in particular whether commercially operated centers should be allowed, has received a lot of attention in Canada. All three types are subject to regulation and scrutiny of their facilities and operations by provinces, and restriction and prohibition of commercial operations exist in several provinces. In 1992, child care centers in Metropolitan Toronto consisted of 42 municipal centers (8.5 per cent), 288 non-profit centers (57.9 per cent), and 167 commercial centers (33.6 per cent). Municipal centers are publicly run, while many of the non-profit centers are run by charitable agencies and many more by community groups (sometimes encouraged by Boards of Education). Some commercial centers are corporations, a few chains are part of large corporate enterprises, but most are run by one person or family and have fewer than five branches. Growth of the municipal sector and growth of subsidized spaces within the private sector are strongly controlled by federal and provincial funding decisions. The municipal government has taken policy initiatives in choosing locations of their centers in the 1970s, and in encouraging the growth of non-profit centers by various methods in the 1980s and 1990s. The future of federal–provincial funding arrangements is in doubt as federal transfer payments to provinces continue to diminish (Friendly, 1992; Friendly et al., 1991). Changes in funding also influence the geography of care: where and how accessible child care spaces are.

The tables also show how many of the centers have purchase-of-service agreements (P.O.S.s) with Metropolitan Toronto; P.O.S.s mean that subsidized as well as unsubsidized children can attend those centers. The municipal centers, owned and operated by Metropolitan Toronto, care for subsidized children only. However, there are relatively few municipal centers, and there has been relatively little growth in this sector. Instead, the share of P.O.S. agreements held by non-profit and commercial centers has grown over time. In 1992 almost 80 per cent of the non-profit and commercial centers had P.O.S. agreements with Metropolitan Toronto, compared to only 58 per cent in 1971. Thus these centers are at least partially dependent on publicly funded clientele.

The bottom rows of Tables 7.1 to 7.3 indicate that the numbers of centers grew rapidly during the 1970s (from 104 in 1971 to 266 in 1981: an increase of 156 per cent). Most of that growth was in the non-profit sector (from 29 in 1971 to 99 in 1981: an increase of 241 per cent), distantly followed by commercial centers (from 55 to 134: an increase of 143 per cent). In each period, the municipal sector was the smallest, and while commercial centers dominated in 1971 and 1981, in 1992 non-profit centers accounted for 58 per cent of Metropolitan Toronto's child care centers. All three sectors grew over the period, although commercial sector growth rates have been generally lower than those in the non-profit sector. Generally, non-profit care in Canada is of better quality than commercial care. A study of 1,000 child care centers in Canada comparing commercial and non-profit centers found that non-profit care was of significantly higher quality than commercial care, probably reflecting lower wages and weaker staff qualifications at commercial centers (Ontario Coalition for Better Day Care,

Table 7.1 Child care centers and spaces in Metropolitan Toronto, 1971

	Total no. of centers	Total no. of spaces	Municipal centers Centers/spaces		Non-profit centers Centers/spaces		Commercial centers Centers/spaces	
Toronto	37	2,113	12	884	16	815	9	414
(P.O.S.)					(10)			(5)
East York	3	138	1	65	0	0	2	73
(P.O.S.)							(1)	
York	6	348	0	0	2	120	4	228
(P.O.S.)					(1)			
North York	15	625	3	165	7	250	5	200
(P.O.S.)					(3)		(3)	
Scarborough	28	1,842	3	160	3	134	22	1,548
(P.O.S.)					(3)		(16)	
Etobicoke	15	708	1	90	1	36	13	582
(P.O.S.)							(7)	
Total	104	5,774	20	1,364	29	1,355	55	3,045
(P.O.S.)					(17)		(32)	

Note: (P.O.S.) = the number of centers with purchase of service agreements: they can accept subsidized children. All municipal centers accept subsidized children.

Source: Community Information Centre of Metropolitan Toronto (1971)

Table 7.2 Child care centers and spaces in Metropolitan Toronto, 1981

	Total no. of centers	Total no. of spaces	Municipal centers Centers/spaces		Non-profit centers Centers/spaces		Commercial centers Centers/spaces	
Toronto	90	4,561	12	937	58	2,693	20	931
(P.O.S.)					(45)		(11)	
East York	12	716	1	65	1	85	10	566
(P.O.S.)							(7)	
York	11	759	2	150	2	167	7	442
(P.O.S.)					(2)		(4)	
North York	62	3,907	8	598	22	1,221	32	2,088
(P.O.S.)					(12)		(20)	
Scarborough	57	3,255	7	460	11	514	39	2,281
(P.O.S.)					(7)		(29)	
Etobicoke	34	2,161	3	267	5	253	26	1,641
(P.O.S.)					(4)		(16)	
Total	266	15,359	33	2,477	99	4,933	134	7,949
(P.O.S.)					(70)		(87)	

Note: (P.O.S.) = the number of centers with purchase of service agreements: they can accept subsidized children. All municipal centers accept subsidized children.

Source: Community Information Centre of Metropolitan Toronto (1981)

Table 7.3 Child care centers and spaces in Metropolitan Toronto, 1992

	Total no. of centers	Total no. of spaces	Municipal centers Centers/spaces		Non-profit centers Centers/spaces		Commercial centers Centers/spaces	
Toronto	155	7,700	14	895	109	5,487	32	1,318
(P.O.S.)					(86)		(17)	
East York	23	1,012	1	35	10	406	12	571
(P.O.S.)					(10)		(11)	
York	31	1,374	3	151	17	687	11	536
(P.O.S.)					(16)		(7)	
North York	128	6,097	10	619	84	3,118	34	2,360
(P.O.S.)					(70)		(29)	
Scarborough	98	5,448	8	442	47	2,155	43	2,851
(P.O.S.)					(31)		(37)	
Etobicoke	62	3,848	6	344	21	1,121	35	2,383
(P.O.S.)					(17)		(30)	
Total	497	25,479	42	2,486	288	12,974	167	1,019
(P.O.S.)					(230)		(131)	

Note: (P.O.S.) = the number of centers with purchase of service agreements: they can accept subsidized children. All municipal centers accept subsidized children.

Source: Community Information Centre of Metropolitan Toronto (1990, plus updates)

1987). Furthermore, while some commercial centers can care for children with subsidies, and have even received start-up and maintenance grants, Metropolitan Toronto tends to favor non-profit centers in its policies and funding patterns.

During the recent recession there has been some growth in the non-profit sector and a decrease in the commercial sector (in 1989, for example, there were 482 child care centers providing full-time care: 42 municipal, 261 non-profit and 179 commercial. This has been the general pattern across Canada.) Also since 1989, the proportion of centers with purchase-of-service agreements has increased from 65 per cent to 78 per cent. In interviews with child care center supervisors in another study (Truelove, 1993a), many center supervisors stated that they had to have the purchase-of-service agreement in order to be open to all, to meet the needs of the community, or to meet the needs of single parents or immigrant families. Many centers valued the agreement more strongly during the recent recession.

The dramatic growth in the number of centers shows some clear intra-urban patterns. For instance, in the 1970s the number of municipal centers in the suburbs grew rapidly. This reflects Metropolitan Toronto's policy of identifying areas of high need for subsidized care and locating municipal centers accordingly. Metropolitan Toronto continues to periodically evaluate the match between services and need (see Community Services Department, 1990, 1992). In addition, each time period indicates a clear pattern of many non-profit centers and few commercial centers in the City of Toronto, and many commercial and few non-profit centers in the other municipalities. In fact, in 1971 and 1981 (with the exception of North York in 1971) commercial centers were the predominant type outside the City of Toronto. These patterns appear to reflect two trends. First, suburbs, relative to the inner city, have traditionally been regarded as higher-income areas that are more likely to attract commercial centers (this is less true today as recent governments have encouraged mixed-income housing in suburban areas). Second, the inner city is the traditional location of services and charitable agencies that might provide child care, and non-profit centers may be tied to the City of Toronto in terms of low rent locations, often in churches. Moreover, there has been a recently, minor trend of growth of workplace child care; these centers are most often in the downtown area, and are almost always non-profit centers (Beach *et al.*, 1993). However, the overwhelming trend from 1971 to 1992 was that 70 per cent of the centers that opened in Metropolitan Toronto were located outside the City of Toronto.

The tables show that three spatial patterns have emerged. Two municipalities – East York and Etobicoke – are dominated by commercial centers in each period, with well over 50 per cent of their child care centers being commercial (this proportion declined over the period). Second, two municipalities – North York and York – were dominated by commercial centers in 1981 (as was York in 1971), but by 1992 non-profit centers had become the most common type, representing more than 50 per cent of the centers. Scarborough is almost in this second category: non-profit centers are now its most common type, but not yet 50 per cent of its centers. A third pattern is exhibited by the City of Toronto which has always had more non-profit centers than any other type. This reflects the commitment of the City of Toronto's municipal government and Board of Education to a variety of policies that strongly favor non-profit centers (see Truelove, Chapter 3, this volume). In all, the distribution of child care centers indicates the strong influence of institutional factors.

TRAVEL TO CHILD CARE CENTERS[3]

The roles of institutions such as non-profit agencies, government departments, Boards of Education, and in some regions of Canada, corporate chains have been extremely important in the development of child care. Through their internal decisions, these institutions have affected the range of choice, quality, price and convenience of child care for many families. The three types of child care centers have distinct geographies in Metropolitan Toronto. This section examines two aspects of this: the distances children travel to centers and the fees their parents pay. Ideally, a study of travel patterns associated with child care should include information on the supply and the utilization of all forms of formal and informal child care (centers, babysitters, nannies and so on). However, this chapter deals only with *center* care, which is the only form of child care with reliable and available information on the supply and the use of the service. A sample of child care centers in Metropolitan Toronto in 1984 was asked for a list of addresses of the children attending their center (excluding those who attend regular school full-time). These data provide six variables:

1 the address of each child attending a child care center full-time;
2 the EA (enumeration area) in which the child lives and its geo-coded location (X, Y of the EA centroid);
3 the center the child attends and its geo-coded location;
4 whether the child's fees are subsidized or not;
5 the weekly price paid; and
6 an estimate of the distance from home to child care center, using Manhattan (city block) distances.

The sample is stratified by type of center and by municipality within Metropolitan Toronto. *Day Care* (described in the previous section) was used to choose the sample. For each municipality within Metropolitan Toronto, every third non-profit center and every third commercial center in the listing were sent a letter (enclosing a stamped, self-addressed envelope) requesting home addresses of their clientele. The necessary follow-up was by telephone and, in some cases, in person. The sample from the municipal (publicly funded) centers was more easily obtained by visiting the Metropolitan Toronto Department of Community Services. Of the respondents, forty-three child care centers provided usable address lists for preschool children using the centers full-time. The geographical distributions of the three center types in the sample are similar to those of the total population of centers in Metropolitan Toronto (although fewer commercial centers responded in North York than was hoped). And the characteristics of the centers in the sample are fairly representative of the population of centers (for instance, in terms of the weekly fees, ratio of qualified to total staff, average capacity and non-municipal centers accepting children with subsidies).

The sample of forty-three centers provided the home locations of 1,619 preschool children who attended the centers full-time. These data provide indirect information on the distance from home to the child care center. In addition, the addresses provide actual home locations and thus permit study of the whole locational context, including whether there are other formal child care facilities nearby. The addresses were also assigned to the enumeration area (EA) in which they are found.[4] The attendance records at the Metropolitan Toronto Department of Community Services for each child care center in the sample also indicate which children are subsidized. Thus the sample could be further divided into subsidized (1,011) and full-fee (608) children. The price of child care varies with the age of the child. For centers with a variety of ages, the price of full-time care for a 3-year-old was used.[5]

Often studies of child care rely on very limited data sets (Kanaroglou and Rhodes, 1990; Rose, 1990; Cromley, 1987; Hodgson and Doyle, 1978; Benito-Alonso and Devaux, 1981; Brown, *et al.*, 1972) or have completely ignored the locational aspects of child care (Weiner, 1978; Robins and Speigelman, 1978; Zigler and Gordon, 1982; Johnson, 1977; Lero, 1981, 1985). Given that this sample of child care centers includes all three types of centers – non-profit, commercial and municipal – the similarities and differences in their locations and their clients' travel patterns can be examined. As the sample covers all of Metropolitan Toronto, a city that is ethnically and economically diverse, a wide variety of children have been included in the sample. Unfortunately, only the ages and addresses of the children, and no other socioeconomic characteristics (such as ethnicity), are available to the researcher. Moreover, the sample, unfortunately, does not provide information about the parents' place of work, or about family structure or demographic characteristics. However, the most important limitation of this data set is that the *demand* for child care centers cannot be analyzed. As the sample does not provide information on the users of other forms of child care, only the utilization, but not the actual demand for child care centers can be measured. So, for example, in this study a latent demand for child care centers could not be measured among parents who wanted their children to attend, but could not find a vacancy or obtain a subsidy. This latent demand is very important, because child care centers provide very few spaces for infants 17 months or younger. Thus, the term used here is "utilization" of child care centers.

Distances traveled to child care centers utilized

The average distance traveled to child care in the sample was 2.9 kilometers (1.8 miles), with a standard deviation of 3.8 kilometers (2.4 miles). Over 80 per cent of the children in the sample traveled 5 kilometers or less from their home to child care, including 22 per cent who traveled less than 0.5 kilometer (see Table 7.4). The sample also identified a few children traveling long distances (probably to a child care center near a parent's workplace). Thus there is a strong distance decay relationship for travel distances (the simple correlation coefficient for the relationship between the log of (Manhattan) distance traveled and the log of the number of children traveling each distance is $r = -0.74$).

Of the forty-three centers in the sample, five are municipal centers which only subsidized children attend, while seven of the non-profit and

Table 7.4 Frequency table of distances traveled from home to child care center, by fee type

Distance (kilometers)	All children (%)	Full fee (%)	Subsidized (%)
0 to 0.5	21.9	16.9	24.8
0.6 to 1.0	18.1	17.9	18.2
1.1 to 2.0	17.8	19.7	16.6
2.1 to 5.0	25.0	25.0	25.0
5.1 to 10.0	11.8	14.8	10.0
10.1 to 15.0	3.6	4.0	3.4
15.1 to 20.0	1.0	0.5	1.3
20.1 to 25.0	0.5	0.7	0.4
More than 25.0	0.4	0.5	0.3
N	1,619	608	1,011
Mean distance	2.9 km	3.1 km	2.8 km
Standard deviation	3.8 km	4.0 km	3.7 km

Note: $\chi^2 = 24.0$ significant at 95%, 99%.

commercial centers in the sample do not accept subsidized children. Thus thirty-one (72 per cent of the sample) of the centers provide care for both subsidized and full-fee children. Statistical tests indicated that the differences in the distances traveled by full-fee and subsidized children are statistically significant.[6] The average distance traveled is slightly shorter for subsidized children than for full-fee children; in particular, a much higher proportion of subsidized children than full-fee children travel very short distances.[7] In part, these travel patterns are the result of the government of Metropolitan Toronto's location strategies for municipal centers, and that four of the five municipal centers in the sample are in public housing complexes (thus families with subsidies are encouraged to attend the center in their building). However, the results also reflect differences in parents' mobility and car ownership (see Mackenzie and Truelove, 1993).

While the sample may overemphasize children who travel shorter distances, Table 7.5 shows that children traveling to municipal centers (all of whom are subsidized) travel shorter distances than children who go to other centers: 37.4 per cent travel less than 0.5 kilometers, which was also the modal category. For non-profit centers, the modal category was split between 2.1 to 5.0 km (22.8 per cent) and 0.6 to 1.0 kilometers (22.7 per cent), closely followed by 0 to 0.5 kilometers (22.1 per cent). This split might reflect two factors. First, many non-profit centers are community based (for example, in elementary schools) and therefore nearby the children's home. In other cases, given that non-profit care is usually of better quality than

Table 7.5 Frequency table of distances traveled from home to child care center, by type of center

Distance (kilometers)	Municipal (%)	Non-profit (%)	Commercial (%)
0 to 0.5	37.4	22.1	16.0
0.6 to 1.0	12.6	22.7	14.1
1.1 to 2.0	22.5	15.6	18.9
2.1 to 5.0	15.3	22.8	31.4
5.1 to 10.0	6.3	12.0	13.5
10.1 to 15.0	4.5	2.3	4.9
15.1 to 20.0	0.9	1.3	0.7
20.1 to 25.0	0.5	0.6	0.3
More than 25.0	0.0	0.6	0.2
N	222	789	608
Mean distance	2.2 km	2.8 km	3.3 km
Standard deviation	3.4 km	4.0 km	3.7 km

Note: χ^2 = 97.0 significant at 95%, 99%.
But table is so sparse that χ^2 = may not be a valid test.

commercial, parents are willing to travel farther to a non-profit program. Of the three types of centers, children attending commercial centers travel the farthest and the modal class was very clearly 2.1 to 5.0 kilometers (31.4 per cent), strongly suggesting that commercial centers are more oriented to families who travel by car. This result is not surprising, given that the spatial patterns of child care centers in Metropolitan Toronto show that commercial centers represent an important component of child care provision in the suburbs. That commercial child care users travel farther might especially be the case if commercial centers are oriented toward higher-income families in suburban locations.[8]

As discussed earlier in this chapter, there are differences among municipalities within Metropolitan Toronto in the number of municipal, non-profit and commercial centers. There are also somewhat lower densities of child care centers in the suburbs, a trend that was more pronounced at the time of this study. It is not so surprising that there are also strong differences among municipalities in distances traveled to child care. Table 7.6 shows the mean travel distances by municipality for families who pay full fees and those with child care subsidies. Notice that travel distances for the two fee levels are the same in Etobicoke and East York (where, as the previous section showed, commercial centers dominate); and differ by only 0.5 km or less in Toronto and York, with full-fee children traveling slightly

Table 7.6 Mean travel distances by municipality for families paying full fees and subsidized fees

| | Full fee | | Subsidized | |
	N	Mean (km)	N	Mean (km)
Toronto	214	2.1	307	2.6
East York	59	3.5	83	3.5
York	42	2.5	74	2.8
North York	54	4.2	206	2.4
Scarborough	180	3.8	223	2.8
Etobicoke	59	3.4	118	3.4
Inner City	214	2.4	307	2.8
Suburbs	394	3.8	704	2.8

shorter distances (the City of Toronto has the densest network of child care centers within Metropolitan Toronto, moreover, full-fee children in the City of Toronto have the shortest mean travel distances of all). Full-fee families travel farther than those with subsidies in Scarborough (1.0 km) and North York (1.8 km). Of course, these results come from small sub-samples: there are only four centers sampled in Etobicoke[9] and eight in North York. At the bottom of Table 7.6 is a comparison between the "inner city" (City of Toronto, York and East York) and the suburbs (the three outer municipalities: Scarborough, North York and Etobicoke). Average travel distances are longer in the suburbs for full-fee children, but there is not an inner city–suburban difference for subsidized children. The lack of difference for subsidized children may be due to Metropolitan Toronto's policy of choosing nearby centers for subsidized children.

Travel beyond the nearest center

It is clear that there is a strong distance decay effect in travel to child care, with children attending child care centers close to their homes. However, it would be useful to consider what proportion of families actually do travel to the nearest child care centers for which they are eligible, and what proportion are bypassing centers nearer their homes. In particular, are families without subsidies traveling farther to reduce fees (subsidized families' child care fees do not vary with the center attended)? And are commercial centers being bypassed in favor of non-profit centers that are seen as better quality? To answer these kinds of questions, the data were reconfigured so that for each of the approximately 3,000 enumeration areas (EA) in Metropolitan Toronto, two sets of "three nearest child care centers" were identified: one for centers that accept children with subsidies and one for centers that care for those paying full fees (so the second set do not include municipal centers because they only accept children with subsidies). The distances were measured as Manhattan (city blocks).

Table 7.4 reported that the mean distance from home to child care center for the total sample of 1,619 children was 2.9 kilometers. However, the reconfigured data show that the mean distance from home to the nearest center is 2.3 kilometers (standard deviation of 4.1 km). Thus most children do *not*, in fact, go to the child care center nearest their home but to one farther away. Among the numerous reasons why children do not attend their nearest child care centers are parental preferences and lack of space available for a child's age group in the nearest center.

The distances from home to the center actually attended were compared to the distances to the nearest center. Given the strong distance decay effect and that young children generally have low tolerance for lengthy travel, distances of less than a kilometer were calculated. The results in Table 7.7 indicate that 48 per cent of the sample attend their nearest center or one at a similar distance (within 100 meters) from home. Subsidized children (51.7 per cent) are more likely than full-fee children (41.1 per cent) to attend their nearest center (or one less than 100 m more distant). There are statistically significant differences for full-fee versus subsidized children, meaning that the two groups have different tendencies to travel beyond their closest child care center. Part of the explanation for the more limited travel patterns of children with subsidies may be due to Metropolitan Toronto encouraging families with subsidies to attend their nearest municipal center. Indeed, when the same analysis was applied to the sample according to type of center (not shown on a table), a much higher proportion of children attending municipal centers (64.4 per cent) go to their nearest center or one at an equivalent distance than for those attending non-profit (47.1 per cent) or commercial (42.4 per cent) centers (again, a test showed that differences

Table 7.7 Difference in distance between nearest center and center attended, by fee type

Difference in distance	Total	Full fee	Subsidized
100 m or less	47.7	41.1	51.7
101–500 m	6.3	6.9	5.9
501 m–1.0 km	7.9	12.0	5.4
1.0–2.0 km	9.3	9.2	9.3
2.1–5.0 km	17.2	17.8	16.9
5.0–10.0 km	7.5	9.1	6.6
More than 10.0 km	4.1	4.0	4.2
N	1,619	608	1,011

Note: $\chi^2 = 33.4$ significant at 95%, 99%.

in the distances are significant for those attending non-profit, commercial and municipal centers).

It is possible that full-fee children travel significantly longer distances than subsidized children because their families use a lower-priced, but more distant center in order to avoid higher-priced nearby centers. This is feasible as there were large price differences among non-profit and commercial centers when this sample was taken. However, for full-fee children there are only small differences in prices between those of the center attended and the nearest center (for instance, no child attended a center more than $10 per week more expensive than their nearest center). Thus, a lower price does not appear to motivate parents to travel farther. Among all the families paying full fees, only 20 per cent of parents whose children did not attend their nearest eligible center paid $5 or less per week for child care than the price at their nearest center (of course, this might reflect the limited availability of lower-priced spaces closer to home). In other words, the cost of care suggests that prices in a neighborhood are similar, because most of those families whose children do not attend their nearest center are paying a very similar price farther away.

However, when the full-fee sample was subdivided into whether commercial or non-profit centers were attended, it became clear that families seem to be willing to save money by attending a non-profit center: 6 per cent of those attending non-profit centers compared to only 12.6 per cent of those attending commercial centers were saving at least $5 per week over the fees at their closest center. Of course, given that in Canada non-profit centers are generally of higher quality than commercial centers, these parents are going to be motivated by issues beyond saving money. Among other reasons why families do not attend their nearest center might be the reputation and quality of centers farther away (like non-profit centers). The nearest center may be full (i.e. the center of preferred "choice" may not be the one used) and it may not serve the child's age group. In addition, as simple Manhattan distances rather than actual road distances (or travel times) were used in this analysis, the center may not actually be as accessible as other centers farther away, or may be only marginally closer than the center attended.

CONCLUSIONS

Geographical analysis can contribute to an understanding of the effects of present and proposed policies on child care, particularly their spatial and redistributive impacts. New government policies (such as a Province of Ontario ruling in the late 1980s that all new school buildings must include a child care center) can be analyzed for their social and spatial impacts on efficiency and equity. This chapter examined the spatial evolution of child care centers in Metropolitan Toronto since 1971. Like Canada in general,

Metropolitan Toronto has a mix of ownership types of child care centers: non-profit and commercial predominate, but municipal centers are also important. Each type has a distinct geography: municipal centers are clustered in the City of Toronto, although they have grown in the suburbs over the last twenty-five years. Non-profit centers can be found through the region and are beginning to take over from commercial centers as the leading provider in the suburbs. This pattern is clearly reflective of the actions of a set of institutions which are important in the provision of child care.

This chapter also examined families' use of, and travel to full-time child care centers. The average distances from home to child care center are relatively short and there is a strong distance decay effect. Subsidized children travel shorter distances on average than do full-fee children, while children attending municipal centers (all of whom are subsidized) have the shortest travel distances of all. These patterns are related, in part, to the policies of the government of Metropolitan Toronto in recommending centers near to a family's home, and building municipal centers near and in public housing complexes. The distances traveled to child care differ by municipality within Metropolitan Toronto, with the shortest distances and densest network of child care centers in the City of Toronto. While most children do not attend their nearest center, even in the suburbs they travel only short distances beyond their nearest center. In terms of prices, the greatest fee savings were made by those attending non-profit centers, but in general, price differentials do not explain travel patterns. Undoubtedly, choice of center is related to availability of spaces, parents' evaluation of centers, convenience to work location, transportation mode and other factors. Unfortunately, data constraints meant that the data set did not include sufficient individual level information to investigate these many factors.

There is now a great opportunity to analyze the impacts of a variety of policies concerning child care. Geographers can make clearer the spatial impacts of policies that initially appear to most people to be aspatial, and can analyze how spatial policies affect well-being. Policies should be evaluated in terms of whether they provide parents with real opportunities and choices in child care; similar evaluations can be carried out for other services. Social scientists need to examine the effects of new policies to consider what alternatives are available in terms of the funding and the provision of a service. Recent proposals for differential government funding of child care (with lesser amounts of money, or none, for commercial centers) require further detailed analysis to ascertain their effects on accessibility, demand and equity.

More research must be focused on the informal sector of child care provision; this sector cares for approximately 80 per cent of children receiving care. Only when formal and informal child care are both considered can the demand for child care be analyzed thoroughly, as these sectors are substitutes for each other. If all forms of child care are examined we can learn a great deal about how people choose their child care, how they weigh each alternative, and what characteristics of each sector are most important. Looking at one sector alone limits our ability to analyze policy issues (Warren *et al.*, 1988). For example, the question of equity could now be framed in terms of how the total supply of an urban service – not just that portion provided by government – is distributed within a community (see Truelove, 1993a).

8

MOTHERS, WIVES, WORKERS
The everyday lives of working mothers
Kim England

The continued trend toward large numbers of mothers in paid employment in the US highlights the pressing need for quality, affordable and accessible child care. In the 1950s gender roles and the division of labor between married, heterosexual couples were clear-cut: men were "breadwinners" and women did the domestic chores and child rearing. Today, women do not view their paid labor force participation as intermittent and since the early 1980s women in the US (especially younger women, including mothers) are more likely to search for full-time rather than part-time employment opportunities. Although over half of all women are in paid employment and contribute an increasingly larger share to their family's income, the division of household responsibilities has not altered proportionately. In this chapter I explore the everyday lives of working mothers in Columbus, Ohio, paying particular attention to the coping strategies they develop to enable them to negotiate their multiple, but not necessarily overlapping, roles.

WORKING MOTHERS, THE PROVISION OF CHILD CARE AND HOUSEHOLD RESPONSIBILITIES

Despite the rapid increase in women entering the paid labor force, the traditional gender role of women as "natural" caretakers of the home and family has remained remarkably unaltered. Male partners of working women do roughly the same amount of housework and parenting as men in "traditional" relationships. In fact, even in heterosexual couples where the woman is employed and the man is unemployed, most domestic responsibilities remain "women's work" (Blau and Ferber, 1992; Bowlby, 1990, Hochschild with Machung, 1989; Michelson, 1988).

Recent evidence suggests that domestic activities, especially parenting, are becoming more equitably divided in heterosexual couples. This seems to be especially true of couples in their 20s and 30s (Blau and Ferber, 1992). And Pratt and Hanson (1993) found a more equitable division of domestic responsibilities among couples where the woman was not employed in a female-dominated occupation. This suggests a shift in attitudes toward more egalitarian gender roles and relations. Bianchi and Spain (1986), for example, note that opinion polls indicate a decreasing number of people see child rearing as the responsibility of just the mother (regardless of her employment status), and that most people claim to believe that child rearing is a joint responsibility. However, women still tend to have prime responsibility for child care, including making child care arrangements, chauffeuring them to and from child care, and taking time off work when they are sick (Michelson, 1985, 1988; Rosenbloom, 1988, 1993).

Although the supply of child care centers is increasing, most working parents' children continue to be cared for in the informal sector. About two-thirds of children in the US are cared for in private homes, either in the child's home or in a caregiver's home, while about one-third are in child care centers (Phillips, 1991; Reeves, 1992; Thorman, 1989). The increase in the paid employment of mothers over the last twenty-five years has sparked debates over the effects of non-parental child care on children. Consider, for example, the stigma attached to the concept of "latch-key children." As working mothers have gained a greater presence in the labor force, attention has focused on increasing the supply of formal child care.

During the 1980s the place of employers in child care provision became an added dimension in the provision of child care. Although still relatively scarce, on-site child care facilities benefit the firm as well as the employees who use them. Firms with on-site provision report enjoying increased profits as a result of reduced absenteeism, lower turnover rates, higher levels of employee retention, increased productivity and stronger employee loyalty. Few firms in the US offer on-site child care, mainly because costs can be prohibitive. The start-up costs alone can be well over $100,000 and the annual expenses may total more than $3,000 per child. Not surprisingly then, it is mainly large firms that provide such a service (Klein 1992; Reeves, 1992; Thorman, 1989). However, increasing numbers of firms offer other forms of child care benefits and services. These include grants to local non-profit groups to establish a child care center near the workplace, flexible personnel policies (including job-sharing and "flex-time"), child care reimbursement and voucher programs, child care information and referral services, and on-site seminars and counseling on parenting (Cromley, 1987; Thorman, 1989). (See Bloom and Steen, this volume, for a fuller discussion of the role of employers in the provision of child care in the US.)

SPACE AND WOMEN'S GENDER IDENTITIES

The increase in the number of working mothers in the US has occurred against a backdrop of cities designed around the separation of land-uses. Many residential suburbs are spatially separated from paid workplaces, so home, paid workplace and child care facilities are frequently in different places. Clearly this complicates the daily life of the parent (usually the mother) who has to take the child(ren) to child care. When women, especially mothers of young children, combine their domestic roles with paid work, they face complex time–space budgeting problems that have to be resolved within a finite period of time. So women, more often than men, have multi-purpose or linked trips that combine commuting with trips to child care and running errands. This usually necessitates the use of a car, as these activities are rarely located in close proximity to one another.

Several time geography and commuting studies have captured the spatial and temporal constraints of working mothers' daily activities. These studies highlight the complex daily schedules women create around the location and hours of child care arrangements and paid work, as well as their mode of transportation (Droogleever Fortuijn and Karsten, 1989; Michelson, 1985, 1988; Palm and Pred, 1978; Pickup, 1984; Tivers, 1988). Compared to men, women are less likely to own or have access to a car, and are less likely to hold a driver's license, making them more dependent on public transport. Furthermore, most studies show that women generally have shorter work-trips than men.[1] This suggests that working women, unlike working men, are more likely to trade off commuting time to accommodate their domestic responsibilities (Hanson and Pratt, 1988, 1991; Johnston-Anumonwo, 1992; McLafferty and Preston, 1991; Michelson, 1988; Wekerle and Rutherford, 1989).

There have been very few geographic studies that specifically address child care. Much of this literature has focused on the increased provision of child care or sociospatial issues surrounding the daily care of children. Mackenzie and Truelove (1993) examine the growth in formal and informal child care in Canada, drawing out the connections with broader economic and political changes and the shifting roles of women. Bowlby (1990) argues for increased child care, but points out that unless the domestic division of labor is modified, more child care will not fundamentally change gender inequalities. Fincher (1993) and D. Rose (1990, 1993) explore changes in state policies regarding the provision of child care. Fincher considers state policies within the nexus of human diversity, illustrating how state policies privilege particular groups of women. Rose pays attention to the ways in which the local manifestation of state policies are shaped by the intersection of family structure, class and ethnocultural origin (also see Rose and Chicoine, 1991).

In terms of the daily care of children, Michelson (1985) notes that about 70 per cent of journeys to child care involve the mother, usually alone (55.6 per cent), but sometimes accompanied by the father (14.8 per cent). He also found that journeys to child care tend to lengthen work-trips by about 28 per cent. Cromley (1987) explores the location problems and preferences of a group of parents using formal child care. Most parents preferred care that was either near home or near their workplace. She found that availability and affordability posed greater problems than finding care at a convenient location. Dyck (1989,

1990) examines the strategies that mothers use to integrate their domestic and employment roles, underscoring that local context is central in their creation of strategies (also see her chapter in this volume).

Recent literature on the geographies of working mothers has highlighted the fluidity of the everyday as women weave together their daily activities and responsibilities. Feminist geographers emphasize the interconnections and linkages between public and private, home and paid workplace. Currently, there is a shift toward underscoring women's transformative capacities. Women are cast as active, knowledgeable strategizing agents, able to plan and learn new ways of acting and possibly change their sociospatial context to better fit their needs. This is not to deny that women face spatial and structural constraints – clearly they do. Rather, feminist geographers are creating approaches to research that mesh the biographies of individual women with the sociospatial structures in which their everyday lives and practices are embedded. And the local context shapes the meaning and diversity of their practices. I pursue that image of women in my exploration of the everyday lives of working mothers in two suburbs of Columbus, Ohio.

BACKGROUND TO THE CASE STUDY

I investigate the everyday lives of working mothers in two suburbs of northern Columbus, Ohio – Westerville and Worthington. Columbus is a medium-sized, fairly compact city. Although Columbus still has a distinct downtown, most urban land-uses have undergone suburbanization since the Second World War. Much of Columbus's population growth and rapid urban development has been in and around my study area. Westerville and Worthington are well-established, very affluent, residential post-war suburbs, and a large portion of the residents are employed in professional and managerial occupations. Until the late 1970s there were few opportunities for local employment. Now there are more diverse land-uses, although office functions predominate and manufacturing is limited.

My discussion is based on the analysis of two sets of in-depth, unstructured, interactive interviews conducted in the late 1980s for a larger project (see England, 1993a, 1993b). The first set of interviews were with twenty-five suburban women with children. I mailed the women material introducing me, outlining my research, and included a list of possible interview items and a sample interview. Later I telephoned the women to arrange to interview them. The majority of the interviews took place in the women's homes, and lasted approximately 3–5 hours.

Two-thirds of the women were married, all are white and they ranged in age from 27 to 64, although most were in their 30s and 40s. Most of them had at least one child aged less than 18 years old. I divided the women into four groups based on the age of their youngest child: seven had children aged less than 6 years old (preschoolers); eight had children aged between 6 and 12 years old (school-aged); five had children aged between 13 and 18 years old (teenagers); and five had adult children. The women with young children provided a picture of current strategies for combining motherhood and paid employment, while the older women with grown children enabled me to gain insights into how working mothers lives have changed over time. The interviews were intended to provide me with an in-depth, rich understanding of the variety of daily coping strategies mothers create, and to explore their efforts to alter their sociospatial context to better cope with their various roles.

The ability of women to actively develop a set of coping strategies and alter their sociospatial context is mediated, in part, by wider changes in societal institutions and practices that enable women to better negotiate multiple roles. One such change is the recent increase in the employer provision of child care services and greater flexibility in personnel policies regarding families (represented by the increased media attention on "family-friendly workplaces"). Thus, the second set of interviews were with ten personnel managers of large offices in northern Columbus that employ sizeable numbers of women. The personnel managers of the firms were contacted and interviewed at their workplaces. Their interviews lasted between 1–2 hours.

TALKING WITH WORKING MOTHERS

In this section my goal is to capture the complexity of the everyday lived reality of working mothers. I draw on the words of the women and the managers whom I interviewed (they are identified by pseudonyms). My discussion draws on the notion that space is not only the context, but the medium through which the women's everyday practices occur, and that those practices are made and re-made in place. In some instances the women draw on localized social relations and networks to create strategies to cope with their various overlapping and often contradictory responsibilities. The first section deals with the transition between home and work, a time that is very stressful for working mothers. Then I explore various strategies adopted by the women for combining mothering and paid work. Finally, I examine the women's strategies around domestic chores and child rearing at home.

"When two worlds collide": making the transitions between home and work

I set the alarm for 6:00 and try and drag everyone else up by 6:15. Paul [her partner] usually makes coffee because I'm running around getting things ready. I get everyone else moving. We try to leave here by 7:20.

113

We never make it. It's usually a mad rush – making sure everyone has keys and glasses and briefcases and all those good things. Then I take Charlie [their 9-year-old son] to school and I drop off Alice [her 19-year-old daughter from her first marriage] at work, then I go to work.

(Jean, aged 46, married (2nd), two children)

In describing the daily routines of two-income couples, Michelson (1988) notes that the most stressful daily activity for women (but not men) is the transition between home and work. At these times the linkages between time and space are very apparent, but the stress is greater for women than men because women, as Jean shows, have more responsibilities during and around these transitions. For example, Michelson found that women were four times more likely to be responsible for trips to the child care center or babysitter. Among the women I interviewed, many of those with partners claimed that child rearing and child care was their most equally shared domestic activity (this was particularly true of the younger women). However, on closer inspection, I found that it was the women who had primary responsibility for taking the children to child care and looking after sick children. "The kids have always called me," said Penny, "I've always been the one to take them to the dentist, the doctor."

Many of the personnel managers were aware of their women employees' "transition tensions," and the stress created when family obligations clash with work responsibilities. Most told me that their firms have become much more flexible about employees taking time off to attend to family needs (reflecting the emergence of the "family-friendly workplace"). Many remarked that this was in direct response to the difficulties they knew their female employees faced in negotiating multiple roles. The personnel managers said that in most cases arrangements are made with department supervisors and, as long as it did not become a chronic problem, employees did not usually face resistance from their managers, and many even found that they did not lose any wages. As one personnel manager explained:

[Absenteeism due to family obligations] can be a problem, if it's not managed appropriately. But it's more in terms of how well the supervisor handles this. Basically we have five sick days which you can take yourself if you are sick, we have two personal days, and then we have vacation days. It's up to the individual supervisor how to deal with each individual case. But it is also up to the employee. It's not just the supervisor's responsibility. [The employees] generally try to schedule something for either around lunch time and take off a couple of hours that way, or at the end of the day or the beginning of the day.

Not all the women interviewed had such accommodating employers, although most agreed that in recent years, employers had become more sensitive to the difficulties of being a working mother. Liz told me about her previous job that she had prior to her second marriage:

> I used to work at a place where there was a grandfather clock outside the boss's office. It chimed at 8 o'clock, so you were made to feel like a criminal if you were five minutes late. I'm a mother of two children. I didn't need someone hounding me to that point. I knew I'd already put in at least two hours getting Brian [her son from her first marriage] ready and to the sitter. I resented it, especially because once I was there I worked hard, and when a quarter 'til 5 rolled around – when I was supposed to quit – I'd never leave before 6, never. I couldn't deal with it.
>
> (Liz, aged 36, married (2nd), two children, aged 7 and 15)

Of course, among the women with partners there were some exceptions to the traditional division of labor in and around transition times. Wendy's partner, Sam, took care of their children every morning, while Joanne emphasized that child care was a joint responsibility:

> The morning hours are Sam's time, they have breakfast together. I don't have to worry about getting anyone ready to go anywhere, I haven't for years. Sam even takes the little one to the sitter.
>
> (Wendy, aged 36, married, two children, aged 4 and 13)

> Rob and I have always shared equal responsibility for retrieving ill youngsters from the day care center or staying home to attend to a sick child. We still have this arrangement today.
>
> (Joanne, aged 37, married, two children, aged 12 and 13)

The mothers of school-aged children raised additional concerns about child care in the summer, when schools closed and many after-school care programs are not available (see Cromley, this volume, for a discussion of after-school child care). Jamie, Jean and Ruth managed with the help of their older children. Ruth explained:

> The thing that bothers me is the summers, because I don't like to see him [Matt, aged 12] by himself in the house for eight hours a day. So far it's worked out because either Janie's [her 21-year-old daughter, a college student] been home or maybe taking a few classes. She's in and out during the day, not home all day everyday, but enough so that it breaks it up for him. And last summer, Robbie [aged 19] started working full-time and getting home early, around 2:30 p.m., so that kind of helps . . . (and) there's the girl next door and then this one down the end of the street. They're both home, pretty much. He's got people he can go to if he has a bad problem. Of course, there's always the telephone to call me at work.
>
> (Ruth, aged 45, married with four children, three at home, aged 12, 19 and 21)

These women's reliance on their older children illustrates an interesting strategy revealed by my interviews. Older children were often important in easing multiple roles and transition tensions. This has received little attention in the literature, presumably because of the assumption that parents closely space the births of their children. However, other family patterns are emerging. There may be large age gaps among children in blended families (as in Jean's case), or where women have additional children after establishing their careers (as in Jamie's case). Moreover, the increase in so-called "boomerang babies" (where young adults return or continue to live with their parents, partly as a result of the decline in real wages since the early 1970s), means that in many households there are adults other than the parents who can be called upon to provide child care.[2]

Eight of the women whom I interviewed were or had been lone mothers. Obviously they did not have partners to rely on. Thus, lone mothers, arguably, faced greater time–space constraints than the women with partners. I asked Sheila, a lone mother, what was the biggest problem she faced as a working mother. She immediately replied:

Scheduling. It's trying to have enough time to satisfy everybody, including myself. There just isn't enough hours in the day. That's my biggest problem. It's not even money right now. Working mothers with a man in the house or somebody else in the house, however you want to put it, wouldn't have near the difficult time I do. It's just having someone else to delegate some of the responsibility to, some of the adult responsibility.

(Sheila, aged 43, divorced with three children, aged 12, 15 and 18)

The lone mothers' strategies often involved other family members or neighbors and friends. Certainly, grandparents, if they lived nearby, frequently played a central role in the web of coping strategies (as they did for women with partners). A couple of the older women whom I talked to were grandmothers and had daughters who relied on them to provide care. Olive's daughter was a lone mother and worked until 9 p.m. week-nights. Olive provided after-school care for her 6-year-old grand-daughter – feeding her, ensuring that she did her homework and preparing her for the next school-day. On the other hand, Lucy, did not have any family living near her, instead she had chosen to work close to home:

Although I don't drop the children off before work, my decision on where to live in relation to the distance from my job was very important. Being a single mother means my time is valuable. I need to be close to home so after-school appointments can be met – haircuts, dentists, car-pooling for sports activities, and so on.

(Lucy, aged 37, divorced with two children, aged 11 and 13)

Other lone mothers clearly depended on a pre-existing and evolving set of localized networks in order to cope. For example, Sally relied on a neighbor to pick up her children if they got sick. Joanne talked of "trading off with neighbors" in terms of chauffeuring children and creating a babysitting circle.

"Working 9 to 5": combining mothering and paid work

My grandmother lived with us, so she looked after my son when I worked. When I had my daughter I quit full-time work and went part-time. My family has been very important in helping us. My grandmother was always active. She's 93 now. I guess I didn't need day care or babysitters or anything like that. So it's always been a different atmosphere for us.

(Shirley, aged 51, two adult children)

Shirley exemplifies two important points regarding combining mothering and paid work. First, the reliance on family members (often grandmothers) to provide informal (often unpaid) child care. Generally speaking, until relatively recently, this was the only viable option for many working mothers. Second, Shirley switched to part-time work after the birth of her second child in the late 1960s. This used to be the classic coping strategy for working mothers. Firms scheduled "mother's hours" so that women could be at home at the same times as their children. However, only four of the women whom I interviewed worked part-time (including one who job-shared). Of these women, Jamie and Lynne were the only ones with preschool or school-aged children. Olive and Marge had adult children. The reasons the other women gave for *not* working part-time included the limited pay and benefits associated with part-time work, few promotion prospects and personal preference for full-time work. At the same time, there was also a lack of available part-time work. For example, few of the firms where I conducted interviews offered opportunities for part-time work, and even among those that did, there were only a limited number of positions. The general consensus was that there are very few jobs that could be satisfactorily completed in part-time hours; and part-time work tended to be biased heavily toward entry-level positions which the personnel managers felt did not attract too many people.

One reason more women with young children have been able to work full-time is increased availability of child care. The women whose youngest children were teenagers or adults were particularly aware of the growth of child care provision. Many had been in paid employment when their children were young and described the difficulties that they formerly faced in their efforts to secure good and reliable child care. For example, Rose told me:

We had to get a private sitter. Luckily, I knew of this woman from my old room-mate from my single days, who took her child to her. [Rose's children] went to the sitter until they were 12. I took them, she used to live [10 miles away], but she was good and I just didn't know anyone else who would take the kids. But it was worth it because she was so good, a friend recommended her and you don't like to leave your baby with a perfect stranger. We didn't have these nurseries like you do now; I think they're terrific. They're in spots close to where you work. They're great.

(Rose, aged 53, divorced with four adult children)

In addition to remarking on the growth in child care provision, Rose's comments highlight a couple of other significant issues. She illustrates the time–space budgeting that working mothers face (the sitter lived 10 miles away, at a location that was neither convenient to Rose's home or workplace). She highlights the importance of local networks and locally embedded knowledge in terms of learning about her sitter through word-of-mouth. Rose also hints at the guilt that the older women often expressed over leaving their children "with a perfect stranger." Joan had felt this "guilt" acutely:

I felt guilty all the time. I missed not being here when they got home from school, that made me feel guilty. I'm their mother, I'm supposed to be here.

(Joan, 51, married, four adult children)

This "guilt" reflects public debates about the effect of non-parental care on children. Some of the women I interviewed had strong opinions about what they considered to be the negative consequences of child care. Jean told me:

Frankly, I don't like the idea of young mothers leaving their babies with other people. I really don't. I think babies need bonding with their mother, and father for that matter, but the mother is the primary caregiver anyway. I think it is very important. I don't know what's going to happen societally when 75 per cent of mothers are leaving their babies with other people.

(Jean, aged 51, married (2nd), two children, aged 9 and 19)

Contrast this with Clare, who is much younger than Rose, Joan and Jean:

I don't get to see Kyle [her son] during the day, but he's at an accredited day care center, and though I still worry about him, I can see the difference his being there is making. He's very confident and outgoing, it's doing him good.

(Clare, aged 27, married with one child, aged 2)

Clare does worry about her son (perhaps she even feels "guilty"), but clearly was not terribly worried about non-parental care. Indeed, she felt the experience of child care was beneficial for Kyle. Obviously, the women with younger children had benefited from the increased availability of child care and the increased recognition of the difficulties (as opposed to stigma) associated with being a working mother. While the younger women certainly appreciated the greater availability of child care facilities, their major concern regarding child care was the role of the employer. Here is Clare again:

> I think the most important thing firms could do for us is to provide us with on-site day care. I think women would be so much more productive if they know their child was being well looked after, and was nearby so they could visit on their breaks.

None of the firms where I interviewed personnel managers offered on-site child care. Many commented that liability insurance for on-site provision made it prohibitive. Most of the personnel managers acknowledged that child care was a recurring problem. Two of the managers made the following remarks:

> We looked into [on-site facilities] and it was quite cost prohibitive. In order for the company to fund something like that, it was just too much money, compared to what we felt like we would be getting out of it.

> We're not very progressive. I wish we were. I really respect the companies that offer some help in the day care. I really think we're behind the times in not offering something along these lines. Whether it's subsidizing it, offering actual facilities, some compensation for it in some manner, whether it's time off or something. You know, I'd like to see more flexibility, maybe, in the scheduling. A lot of companies have done that, and that helps a lot of people, particularly mothers, in getting the kid off to the day care or the baby-sitter.

The second remark reflects the increased sensitivity of employers to the plight of working mothers. While this was true of most managers whom I interviewed, this particular manager was the only woman manager I spoke to. She was also the mother of a young child and thus had an experiential understanding of the difficulties faced by other mothers employed by the firm.

Although none of the firms had on-site child care, most of the personnel managers told me that their firms had introduced, or were planning to introduce other forms of child care service. "We need to move with the times," said one manager. They felt they needed to react to shifts in labor market patterns and respond to the growing number of mothers on their pay-roll. A few of the firms were experimenting with less cost-prohibitive alternatives to on-site child care. For example, two firms had recently begun to issue their employees with vouchers for nearby child care centers, and a number offered a child care option in their benefits packages.

> We have one benefit that we call a flexible benefit, which you get on a pre-tax basis. You can set aside extra money – before the government taxes it – to cover things which aren't covered, and you can put in as much or as little as you want, within certain rules. Some people put in for their child care this way, and others take the vouchers we offer for the child care center here in [the office park].

However, one of the personnel managers described meeting resistance from both management and some employees over his efforts to introduce child care benefits.

> There's a day-care center right next door. I even tried to have some arrangement with them for our employees. And the road-block I kept running into was that unless we can offer a benefit to everyone that will benefit the majority, we can't have it. . . . But it's brought up all the time, "When are we going to have day care or something?" So I guess we will eventually. Right now we've got other priorities, but I hope for it some day.

This manager's frustration was compounded by his knowledge that the provision of some sort of child care has its advantages for employers. For instance, a number of the personnel managers remarked that their employees who were mothers, especially the ones who were divorced, were among their better employees in terms of having lower rates of turnover and absenteeism.

"Coming home to another job": child rearing and domestic responsibilities

> Since I've started working full-time there's not enough time to do everything. Obviously you can't work forty hours a week and maintain things like they were before you had kids. I think if you have children it takes up more time. And probably most women if they work and they are married, most women still have the responsibility of the home. No matter how much you can talk about equality, it's "I'll help you, but you've got to do most of it." You have to say you do this and I'll do that. You delegate it. So you still have the responsibility.
>
> (Yvonne, aged 28, married, two children, aged 5 and 8)

Yvonne and many of the women whom I interviewed substantiated Hochschild's (1989) conclusion that daily practices have yet to catch up with

120

the beliefs of sharing domestic labor professed in opinion polls. And, as Yvonne remarked, women are still the primary managers of the home. In addition, the domestic tasks that the younger men were most likely to undertake (yard-work and home repairs, for example) still had a traditional masculine tinge to them. Even with child care, the men were more likely to do "fun," occasional activities (trips to the park or zoo) than the more routine maintenance, daily tasks (bathing and feeding). Tammy illustrates this:

> I probably do 60 per cent [of the domestic chores], he does 40 per cent. He cuts the grass, and usually runs the vacuum. I usually do all the cooking. I usually do most of the laundry and the dishes. Although he's pretty good about doing that too. He plays with the kids, but as far as baths and things like that, that's usually me. But he helps quite a bit. It's got better over the years.
>
> (Tammy, aged 26, married with two children, aged 1 and 4)

Those women who felt that their partners did a great number of domestic chores, especially in comparison with their own fathers, were usually among younger women. However, most of the married women I interviewed, regardless of their age, felt that they did more than their "fair share." (Note that even Tammy, who felt she had an egalitarian marriage, talks of her partner "helping" not "sharing.") The women used numerous strategies to cope with domestic responsibilities. These included using a diaper service and eating out at restaurants; as Christine commented: "When we can't be bothered to cook, we just go through the drive-through at Taco Bell." Sally, Yvonne and Sylvia hired domestic help. Most of the women talked of lowering their expectations by becoming less fanatical about having, as Lesley put it, "a *Metropolitan Home* house, when I see dust now, I just leave it. Other things are more important." A few of the women, like Kathy, had actively negotiated a more equitable division of labor with their partners.

> I was doing everything. I thought that I should be "Supermom." I didn't realize that I had the right to ask my husband it's time for you to take the kids out, do the dishes and so on. Then five years ago I decided I couldn't handle it any more. . . . Now he does more than half. We got counseling, because I said if this is what marriage is about, then I've had enough. We decided we didn't want to break up the marriage, so we got counseling and got it all ironed out. He's stuck to his bargain and I've stuck to mine. Whoever gets home first starts dinner. Whoever is in the basement puts a load of laundry in. It's no longer "Mom's job."
>
> (Kathy, aged 33, married with two children, aged 5 and 7)

Just as older children could ease the strains of transition times, many of the women talked of their older children "helping around the house." For instance, Marion's three adult children (aged between 18 and 24, living at

...dest being an example of the "boomerang babies", having
...ome after college) had been assigned domestic tasks for at least
...ious eight years. Marion expected these tasks to be completed
...they went to bed. Her view was: "You eat, therefore you cook and
... dishes. You wear clothes, therefore you wash them and sew on
...tons." A number of the lone mothers, in particular, emphasized the
importance of their older children as a part of their coping strategies. In
addition to playing a crucial role in maintaining the household, the children
of some lone mothers even made a direct or indirect contribution to the
family's income. For example, Sheila and Sally's teenaged children had
taken part-time jobs partly to reduce the economic burden they placed
upon their mothers. In fact, Sheila told me her son often paid for the
groceries, while Audrey said that before she remarried, her son regularly
paid the utility bills.

CONCLUSION

Women in paid employment continue to have primary responsibility for
child care and managing the home. They create various coping strategies
and attempt to alter their sociospatial systems so as to better negotiate their
multiple roles. I explored these strategies based of an analysis of
unstructured, interactive interviews with working mothers and personnel
managers of firms employing large numbers of women. I explored coping
strategies in terms of the transition between home and work, combining
mothering and paid work, and domestic chores and child rearing at home. I
specifically explored coping strategies associated with domestic labor, child
care arrangements, the role of older children and the role of employers in
easing the strains associated with combining motherhood and paid work.
Rather than emphasize the spatial and structural constraints at the expense
of human agency, I have highlighted the women's transformative capacities
in their daily negotiations of the web of sociospatial relations and structures
that shape their everyday practices.

The availability of child care options has grown in the last twenty-five
years. Talking to women of different ages indicated that while care by the
child's relatives, especially grandmothers and siblings was important, center
care has become more widely available. However, the interviews also
indicate that women's role as paid worker is often merely grafted onto their
domestic roles. They continue to have major responsibility for running the
home and arranging child care, as well as chauffeuring children to the child
care center or babysitter. Obviously more child care options ease some of
the tensions associated with combining mothering with paid employment.
However, this needs to occur in tandem with policies and daily
arrangements that challenge the assumption that child care is women's
responsibility.

9

MOTHER OR WORKER?

Women's support networks, local knowledge and informal child care strategies

Isabel Dyck

A diverse body of geographic work on women's spatial behavior and its relationship to the demands of combining waged labor and domestic responsibilities has developed with the widespread paid employment of women with young children. How child care arrangements fit into this overall picture of women's geographies has been approached from different perspectives, most notably with a concern for the equitable provision of child care services, and the reduction of logistical difficulties faced by women as they commute to work. However, the agenda of study has broadened as conventional spatial constraint models informing early research have been rethought. Initial descriptions of women's spatial activity, as they juggled the responsibilities of multiple roles in the context of urban land-use patterns predicated on the physical and social separation of home and work, tended to obscure the human agency of women as they responded to economic and social change. More recently, empirical work has shown women as active shapers of their social and geographical worlds, and has also challenged an uncritical acceptance of such analytic dichotomies as the public and the private, work and home, which have been important in informing analyses of women's spatial activity. This chapter offers a perspective on child care which focuses on the informal, culturally sanctioned solutions to child care needs that women create as they respond to the rapid changes of economic and social restructuring which cast them as wage workers as well as mothers. The approach takes particular account of the active agency of mothers, rather than emphasizing spatial constraints, and draws on their own interpretations of their child care strategies.

The study reported in this chapter investigated the everyday lives of women in a residential suburb of Vancouver, British Columbia, in Canada. The study did not specifically set out to study child care issues, but it was evident that managing the care of their children resulted in considerable conflict for those women who made the decision to engage in wage labor. The interpretation and resolution of this conflict, however, was an issue of considerable complexity which can only be understood in part by focusing on the time–space distribution of child care and workplaces and the juggling of multiple responsibilities. Although location and cost of child care was of importance to the women, the dilemma they faced was framed within a web of local and wider relationships which impinged on how the problem of child care was interpreted and the solutions the women reached. I will show that the care strategies developed by the women in the absence of widespread provision of public child care had a moral dimension concerning the status of the women as "good" mothers. Furthermore, the solutions they reached, in the form of "safe space" for their children, were rooted in their work in the domestic workplace of home and neighborhood. In the first part of the chapter I briefly review the literature which framed the study, I then present empirical findings from the research and discuss the implications of these for understanding the links between the experiential aspects of child care and the wider context of the problem.

BRINGING MOTHERS AND CHILDREN IN

Socioeconomic changes throughout industrial society have had considerable impact on women's activities, and particularly on the nature of their labor. Increasing participation of women in the wage labor force in both the US and Canada has included the incorporation of about 60 per cent of those women who have children under the age of 6. In Canada, a report on the meager provision of public child care and women's reliance on "in home" informal child care (Canada, 1986) called attention to the need for more knowledge about women's use of private solutions to gaps in public provision, and a framework within which this might be approached. With a continuing record of inadequate public provision of child care in other western countries, too, the investigation of how women manage child care concerns in a variety of ways in specific contexts remains on the agenda of feminist geography. Although the locational aspects of child care provision have been examined, as have the ways in which the presence of children shape women's use of the urban environment, less attention has been given to the ways in which cultural notions of motherhood and women's identities as mothers enter into the strategies they adopt. Anyon (1983) pointed out that women face the paradox of a social contradiction in beliefs which simultaneously emphasize the importance of women caring for their young children at home, yet ascribe social status through success in the non-domestic world of paid work. The geographical aspects of such a contradiction are important to consider if we are to gain a greater understanding of the links between the specificity of the conditions of local areas and the ways in which this paradox is responded to.

The majority of work in geography concerned with women's activity has not been framed explicitly within this paradox, but it is within a considerable body of literature interested in women's mobility and activity patterns that women as mothers have been identified as particularly constrained users of the urban environment.[1] Tivers (1985), in particular, recognized the importance of the presence of young children in delimiting women's work and leisure participation, although the socioeconomic position of the individual women would determine the actual influence. More recent studies have elaborated on the ways the specificity of household type, age of children and also "race," mediate women's spatial behavior, as these interweave with the location of labor markets, child care responsibilities and the absence or presence of help of partners in balancing employment and domestic work (Johnston-Anumonwo, 1992; McLafferty and Preston, 1991; Rosenbloom, 1993; Rutherford and Wekerle, 1989). Women as mothers have become increasingly visible in this literature, with children's needs significant in the constitution of women's personal geographies. An increasing interest in the diversity of women's experiences has also furthered analysis of women's experience and use of the urban landscape. Empirical investigation of women's paid labor force participation in different geographical areas has destabilized the dichotomy between home and waged work and located women's child care options and strategies within a wider context of economic and social restructuring (Bondi, 1992; McDowell, 1993). Such studies increasingly reveal the complexity of the interrelationships between domestic arrangements, labor market organization and women's spatial behavior (England, 1993a; Hanson and Pratt, 1988). Although inadequacies in the public provision of child care for employed women has been the predominant focus of geographical work interested in women's labor force participation, there has been a growing interest in the informal support systems that women create to bridge the gaps in government and commercially provided care (Bowlby, 1990; Lowe and Gregson, 1989). The few studies in this area suggest that women are not merely passive receivers of the environment and societal changes, but creatively seek solutions to the competing demands of different types of work. The reallocation of care through the use of home-based child care providers indicates a division of labor among women themselves according to socioeconomic position (Fodor, 1978), and an attempt to adjust time and space in response to economic restructuring (Mackenzie, 1989). There is also some evidence of work sequencing on the part of husband and wife (Pratt and Hanson, 1991b) as a family-household strategy in meeting child care needs. Informal solutions to child care needs, and particularly those involving neighborhood support systems, however, remain understudied. Tietjen (1985) notes that although there is evidence that support networks exist between women, there is a lack of knowledge of the conditions under which such support systems are developed. Her study of mothers in Sweden suggested that children, rather than merely acting as a constraint on women's activities, may, in fact, be avenues for the activation of support:

Children may make demands and compete with network members for mothers' time and energy but they also may provide openings for contact with others and may themselves be sources of support. No research has yet addressed these questions

(Tietjen 1985: 490)

The opportunities that mothers create through child-centered activities are an important aspect of the informal child care solutions developed by the women in the study reported here. In the rest of the chapter, I turn to data from this study, focusing on the conditions underpinning the management strategies adopted by the women as they responded to the practical and moral dilemmas of combining paid labor with motherhood. The women's identities as mothers were central to how they interpreted their lives, and the part played by cultural notions about motherhood is integrated into the account of their child care strategies. The women's strategies in allocating care of their children over time and space is embedded in the notion of safe space, which is defined and negotiated by the women in the course of their daily mothering work in the domestic workplace of the neighborhood. The approach taken does not view the constitution of such safe space as merely "local"; the women's organization and interpretation of their mothering work cannot be divorced from a wider context of a gender division of labor within the wage labor force as well as between paid and domestic labor.

The study draws upon interpretive and feminist work in geography that suggests that while wider socioeconomic forces, including patriarchal relations, structure everyday social practices, it is through the experiences of living in unique forms of general relationships that people generate knowledge and meaning is given to the practices of everyday life (Barnes, 1987; Bowlby, 1986; Gregory, 1981; Sayer, 1992). Thus, particular distributions of job opportunities, housing and services, and sets of neighborhood and work relations all are involved in how motherhood will be interpreted and practiced in a particular locality. In the attempt to link everyday experience with environmental structuring, emphasis is placed on the meaning of action from the perspective of those studied, in addition to investigating the daily routines of social life. While the particular form taken by wider sets of relations gives meaning to the interpretation and actions of mothering work, they are also informed by a predominant cultural model of child rearing prescribing the care of young children by the mother at home.

THE STUDY METHODS

In order to study the conditions under which support networks emerge, including the part played by meanings of the mother–child relationship to women, methods are required that allow the researcher to get close to the

126

everyday interaction of social life. In the larger research project from which the data presented are drawn, ethnographic methods were used to reveal the organization of a combination of mothering work and paid employment, and the meaning these different labor activities had for the women in question. The methods were also guided by an ongoing debate over the researcher–researched relationship, and its connection with the social production of knowledge, that, in feminist scholarship in particular, has increasingly advocated a co-researcher situation so that the researched's perspectives are central to the interpretation presented in the final account of a research project.[2] How women made sense of their day-to-day lives was important in understanding the cultural shaping of their child care decisions. A two-year period of participant observation, consisting of talking with, and observing mothers of young children in a variety of informal interactions on neighborhood streets, preschool and school settings and in the context of extracurricular activities, was followed by semi-structured, face-to-face interviews with a smaller number of women.[3] These women were also asked to complete time–space diaries by logging their daily activity, including when and where they went, who accompanied them, the purpose of the trip and their general comments on the day. Documents and academic literature about the history and current demographic status of the locality provided data in which to contextualize the women's actions and the meanings they brought to these.

In total twenty-five women were interviewed. Initial contacts were made through a parenting class and a seminar series held for mothers with young children. From these the "snowballing" technique of subject selection was used, women referring friends and neighbors who were willing to participate in the study. Only one woman joined the study in response to an item about the research in a local school's newsletter. The research participants, therefore, were self-selected on the basis of availability and willingness to talk about their experiences as mothers. The intention was not to provide a representative sample, but to explore women's experiences in some depth in order to gain a conceptual understanding of their mothering work. One result of this type of recruitment was that the women were all White, reflecting the predominant demographic composition of the area, so that the experiences of "minority" women are not included. All of the women were living in two-parent families with elementary school age children or younger. All had engaged in paid employment for some period since the birth of their children, but at the time of the interview were in a variety of situations which included full-time mothering, volunteer work, part-time and full-time employment. Their socioeconomic status varied but, as Boulton (1983) notes, women's housing and social conditions, including security of income, community facilities and their identification with domestic activity, are likely to be more pertinent to their experiences as mothers rather than the more conventional indicator of class as measured by husband's or father's occupation. Apart from one woman who lived in a low-income townhouse complex, the women of the study lived in single family dwellings. Their homes varied in age, design, size and relative location to services, but all were located on streets where two-parent families were the norm. The women had the potential use of a wide range of amenities and services to do with the everyday provisioning of families, and all had some access to a car for transportation to the many recreational and educational facilities provided for children in their particular municipalities.

THE WOMEN'S WORK EXPERIENCES

The locality providing the setting of the study constitutes a suburban subregion of Vancouver. Over half of the women interviewed were migrants or immigrants (from Britain, Europe, Australia and South Africa) and most came to the study area during the 1970s as married women with small children, or anticipating motherhood. Understood as a "family place" by the women, the area features predominantly single family dwellings, numerous commercial and personal services, recreational facilities and considerable park and open space. While the desire to "stay at home with the kids" was a common theme from the women's accounts of their lives, in fact, few had retained a stable relationship to the wage labor force since having children (see Table 9.1). A clear dichotomy between homemakers and wage workers was not useful in understanding the women's experiences, for a general picture emerged of women "dipping" in and out of paid employment according to a number of strategies developed by the women in balancing individual career development, contributions to the household income and the responsibilities of child rearing. Such strategies included part-time employment and frequently involved short-term or interrupted involvement in paid labor. For example, fifteen of the women had engaged in temporary, part-time employment at some point during their marriage in an attempt to supplement the main family wage without disrupting their household and child rearing activities. Five had helped with family businesses for varying periods of time, while at the time of the study five others were in full-time paid employment outside the home.

Those women who had re-entered the labor force after being at home for a period with their children, had done so by different means, including career change. Some had upgraded their qualifications while child rearing, while others saw themselves as distinctly disadvantaged in the job market on finding that their qualifications no longer had currency, and they had to take less skilled work or consider retraining. While transitions between different types of labor made by the women were interpreted by them as choices related to meeting family needs or personal career development, such personal decisions were also shaped by local labor markets and wider

Table 9.1 Occupations of women interviewed in suburban Vancouver (numbers of women in brackets)

The women's occupations at marriage

Accountant (uncertified)
Assistant bank manager
Book-keeper (4)
Clerical worker (6)
Hotel receptionist
Laboratory technician
Assistant personnel manager
Nurse (2)
Office supervisor
Physiotherapist
Secretary (2)
Teacher (3)

Current occupations outside the home

Book-keeper (2)
Middle manager
Nurse
Personal assistant
Real estate agent
Secretary (2)
Teacher
Technician – financial services

Note: The above occupations refer to the time of the interview. Several women were anticipating returning to work in the near future, had recently left a paid job or were "re-tooling" to increase their marketable skills. This list does not include the volunteer work that women engage in and which varies widely in terms of commitment.

economic trends which reinforce women's position in society as secondary wage earners.

The range of local employment opportunities available to the women reflected the steady growth of commercial development in the area, tending to rest in the lower paid, part-time sales and service occupations characteristic of much female employment more generally. Medical services have consistently been a major source of women's employment in the locality, with retail sales and a variety of clerical work in banks and various professional, business and public service offices providing further common sources of employment (Chamber of Commerce, 1986). One aspect of economic restructuring in the region has been a clustering of higher-income quaternary occupations in the city of Vancouver (Ley, 1980; North and Hardwick, 1992), making commuting a necessity for most women seeking senior white-collar jobs. This was the case for two women in the study, while others commuted to inner suburbs or worked within the study area. The particular employment experiences of the women reflected local labor market opportunities and were consistent with national employment patterns. The predominant employment of women in clerical, service and health occupations forms a gender-differentiated part of the labor market which, in general, pays lower wages and has fewer opportunities for career advancement for women than do occupational categories of male employment (Labour Canada, 1986). Part-time work is a further dimension of the gender differentiation of the labor force, this being a predominantly female pattern of employment. In 1991 part-time work accounted for 18 per cent of British Columbia's total employment.

Public policy and economic change also frame women's work experiences. The organization of social services and health care (Doyal and Elston, 1986), for example, and the structuring of mothering work by the school (Smith, 1986) reflect and are predicated on the understandings that women take primary responsibility for household and mothering work within the division of labor. The extent and distribution in time and space of public child care provision in Canada also enters into women's labor choices, in that it reaffirms a gender differentiation of domestic labor which is simultaneously upheld by cultural norms and the values of an "expert" discourse that advocates that children should be looked after by the mother at home (David, 1984; Jaggar, 1983; Rose, 1993). Certainly, the accounts of the women in the study, whatever the current combination of domestic and wage labor engaged in, indicated a clear identification with the domestic sphere of activity. Care of the child was a prevailing concern for the women, and their decision to spend at least some time at home full-time with their children, was expressed as an active choice on their part. The ability to adhere to this choice, however, was difficult as a surge of economic growth in the 1970s was followed by global recession which impinged on both men's and women's patterns of work.

The effects of this recession reached workers in many occupational categories due to a combination of automation, a provincial government restraint program, in which downsizing particularly affected the public service sector, and a multiplier effect in the economic downturn that had ramifications throughout the different economic sectors (Magnusson *et al.*, 1984). Major budget cuts were experienced in the forestry service, funding for research and development, public schooling, post-secondary education, child care, vocational training and health care services (Marchak, 1984).

Of the women interviewed, almost half experienced the impact of recession directly through their husband's unemployment or reduced wages, while others commented on the reduced spending power that accompanied a static income. Changes occurred in the women's own working conditions whether as mothers or wage-workers. One lost her job, one experienced reduced hours, whereas some continued with paid employment instead of following an earlier decision to engage primarily in mothering work, and others entered the wage labor force earlier than anticipated. In a few cases household work was reorganized with an unemployed husband looking after the children while their mother worked outside the home. Some recovery from recession had occurred by the time of the interviews, but in several cases there was a continuing reliance on what had formerly been regarded as "extra" income. Whatever the particular situation of each woman, appropriate care of their children remained a high priority, but participation in paid employment impinged on the manner in which this care could be carried out. While local manifestations of wider economic change affected the women's decisions to work, the impact of these decisions was shaped by general values about "family life" and the women's stories indicated the continuing importance of motherhood to them as they managed the contradictory demands of child rearing and waged work.

RESPONDING TO THE SOCIAL CONTRADICTION

Prevalent values that extol the virtues of motherhood while at the same time devaluing the economic and social contribution of mothering work are part of the context within which women's lives are structured (Dally, 1982). For the women of the study such general societal values concerning women's work contributions and motherhood became contradictory as they responded to the particular social and economic conditions of the locality, and child rearing styles were adapted. Conflict between child rearing and contribution to a family income or career development was reflected in accounts of stress from those women combining different types of labor. One woman who had returned to her previous job as an assistant bank manager to cover for another worker's sick leave commented: "But was it really worth it? With young kids? It's a high pressure, stressed job and it's not nine-to-five and you walk out of it. . . . At the end of six weeks I was a basket case." Another, working on the family farm, said: "I felt everyone was demanding so much of me. I didn't resent it, it was just I was so stressed. It was hard because I still wanted to be a good mother."

Concern over remaining a good mother was a prevalent theme in the women's accounts of decisions to enter paid employment, and worry and feelings of guilt surrounded specific incidents, such as a child's sickness or a special occasion in a child's life. For instance, Anna[4] said:

Probably mothers that are working miss out on the little things kids do. I think they're important to you and you'll remember them when you're older. I remember the first time my son went swimming from the school. On the first day they were going he really wanted me to

131

drive – parents volunteer to drive – it was important to him, but I couldn't do it. Stuff like that.

Responses to these types of conflict and stress were shaped by the women's experiences in the particular settings where women met with their children, discussed common concerns and ways of working out conflicting demands.[5]

For the women of this study the common occupation of mothering in the domestic workplace of home and neighborhood, separated in time and space from wage work activities, comprised a basic constituent of their daily experiences. However, the ubiquitous nature of child rearing and what mothers actually do in rearing and caring for their children has been subjected to little exploration in the literature (David, 1984). The centrality of such work to the forging of support networks was discovered in this research through the identification of different aspects of mothering work, and the manner in which this was organized. Although the physical care of children, particularly those of preschool age, involved considerable time and energy on a women's part, mothering work also involved a range of activity concerned with nurturing the relationship between mother and child, and providing recreational and educational opportunities for the child. Much of this activity occurred beyond the confines of the home and street and included a wide range of organized classes designed for preschoolers, or as extracurricular recreation for school-age children. Such activities formed an area of discretion for parents, but decisions to take part in them were shaped by local norms, what educational and social opportunities public schools provided, and a professional discourse on child development.

During the course of the daily coordination of children's activities women met with other mothers, so creating opportunities for negotiating common understandings about "appropriate" ways of carrying out mothering responsibilities and building situations of reciprocity, such as car pooling or looking after someone else's child for a few hours. In this way, trusted relationships with other mothers, seen to hold similar values concerning the care of children, were forged (Dyck, 1989). It was, therefore, through promotion of the social and educational development of their children during the period when women were at home full-time, that the women began to forge linkages and understandings with other mothers that could be translated into practical and moral support. It was to these trusted relationships that women turned as they extended their activities beyond the domestic workplace, making transitions in and out of the wage labor force. Valerie, talking about what she termed her "supportive neighbor-hood," described how the social dilemma accompanying a woman's decision to engage in paid employment can be allayed through both practical assistance and a discourse between women which imparts moral support – through both positive evaluation of her actions and the sharing of experiences:

> If I want to know anything, or occasionally if I'm wondering about something with my children, I'll pick the phone. It's a very resourceful group of people that I know and it's through the original group I first knew and met when I moved into the subdivision. Somebody will always know somebody who does something, no matter whether it's fixing your stove, doing your landscaping, needing daycare. It's a way of helping people cope with what you've had to cope with. I think we're able to support each other a lot. . . . And I think that's what keeps us going, I really do.

The primary concern of the women as they engaged in paid employment was to provide safekeeping for their children, but how this was interpreted related to a woman's own values and her perceptions of how her choice of child care met the development needs of her child or children in her absence. Through the social links women develop, not only could the care of a woman's children be reallocated in space but the women were assured that the care given was appropriate to their own interpretation of good mothering, and as close as possible to home conditions. In the next section, I present data indicating the experiences of women in creating care situations for their children. These are based on their own accounts and illustrate that although the particular form of their strategies differ, all are underpinned by knowledge acquired in the course of everyday routines of child rearing.

REPLICATING HOME CONDITIONS

In making child care arrangements for their children, all of the women interviewed had, in some way, used neighborhood resources in attempting to gain the control and flexibility needed for their particular family strategy of combining child rearing with a viable family income. Significantly, a combination of resources was drawn upon, and even women in full-time employment rarely relied on only one type of child care source. Most of the women made decisions to take paid employment before arranging child care alternatives. Problems in finding care were not anticipated by women taking part-time employment as flexible or short work hours could often be accommodated within existing networks. Anna, for example, working full-time as a book-keeper in a temporary, summer position, depended on existing neighborhood contacts or kin for child care. Her husband, a truckdriver, looked after the children for most of the time, using his five weeks holiday and banked overtime. Further care was provided through a school-based contact and kin:

And then I just hired our regular babysitter. . . . He comes from a large family. He's the second oldest, and his brother is in Grade 1 with my son. And so I just got talking to, you know how you get talking while you're waiting for the kids to come out from school – so she can keep us in babysitters for along time! And my niece, she's off and on from school, so she was here part of the time. They weren't like strangers. They're people we know.

Such informal arrangements may be difficult to maintain for sustained periods of time, and particularly for women who enter full-time employment and need regular arrangements to fit around their work schedules. Nevertheless, the emphasis on already knowing the sitter and an attempt to replicate home conditions was common in women's arrangements. The reliance on people the women know assures them that their children will not only be cared for in material terms, but will be looked after in a socially and emotionally appropriate environment. In talking to the women, criteria for a safe space in which to leave their children without worry emerged. The qualities making up the different dimensions of a child's well-being were described by the women in terms of physical safekeeping, emotional nurturance and opportunities for academic and social development. Although the specific components of these qualities varied according to individual preferences and perceptions, in general the desire was for "a home away from home." Alison described her preferred choice in the following way: "They're talking about a daycare opening at the new church, but I wouldn't want that anyway. I prefer more – like a family – with a mom and dad. Preferably a friend with children the same age as mine."

A desire for a good relationship between the caregiver and the child was also reflected in other women's comments. The use of kin terms occurred in one woman's description of the caregiver as "really good to the kids; they really like her, they call her Aunty Linda." Another described the caregiver as "an extremely caring lady, and [she] runs the business around the children's needs rather than fitting them into her day." And Deirdre, as she described her daughter's first day at child care, went into some detail about what she valued about the situation:

> She [the caregiver] had a tea-party all set up and ready to go, waiting for Carrie to come. They took Carrie into their home just like their daughter. If we were a bit late – we've always been good at phoning if we're going to be late and ask the babysitter if it's O.K. – she had her children enrolled in a lot of activities after school and she would say, "no problem," and she'd pack her up and take her along with them and then drop her off at home. She'd go out shopping with them, the whole thing. It was really good daycare. Carrie learned a lot there. I worked a lot of day shifts and [she] spent a lot of time with them doing activities, a lot of fun things with them, that were actually good learning experiences.

While the women were clear about what they wanted in the ideal child care situation, access was less certain. In the next section the cases of a number of women are presented to illustrate something of the range of experiences in creating a current "best" solution.

"WORKING IT OUT"

The strategies used to gain access to what the women considered good care, commonly relied on locally based social networks, although a period of "trial and error" was sometimes involved. Instability in care arrangements was also experienced as the situations of caregivers (themselves women with young children involved in their own strategies of combining earning income and child rearing) changed, leaving women with the task of making new arrangements. Attempts to create stability, and so reduce stress, are evident in all the following cases. Keeping the child linked into neighborhood-based social relationships was also a constituent part of the strategies used. Remaining close to already established friends in the neighborhood was viewed by the women as important to the child's emotional well-being and, in addition to being a convenient arrangement, carried the assurance of "known" local values and ways of doing things.

Mary's case is one showing transitions in care arrangements over time, as her own situation and those of caregivers changed. At the time of the interview, her mother, who lived close to the school where Mary taught two days a week, looked after Mary's youngest child. Mary felt that she and her mother had similar ideas on bringing up children and both were, as she said "very, very family oriented." Her oldest child, a boy, who attended kindergarten, went to a neighbor after school, someone Mary has known well since their children were in preschool together. She had wanted someone close by so that her son could go to the local school with his neighborhood friends. However, when her children were younger and Mary was working every morning to supplement the family income during the recession, she had experienced less stability in finding care for her children. This had added to the stress of what she described as "doing too much" in trying to combine her teaching and home life in a way that satisfied her wish to be a conscientious worker, and retain a strong focus on family relations. Her children were aged 6 months and nearly 2 years, and although her job was classified as part-time, she found that it often demanded close to full-time attention. The extra school work she brought home was accomplished first by "trading off" (reciprocating care of children on an occasional basis) with other mothers in the neighborhood, and later by completing it while one child was sleeping and the other at preschool. Her personal contacts had been effective in finding care for her children, but the changing situations of the women providing the care also resulted in stress for Mary when she had to search for new care sources, as she recounts:

135

I got a woman who I met through a post-natal class I went to, and she had a little boy the same age, and she was willing to take on a child. It was just super until she decided to go back to work four months later. . . . I was just devastated. Here was my baby and I had to look for someone new. I'd had someone so perfect – I was in tears, the whole bit. . . . I phoned the Health Unit and I got a list of licensed daycares – and so I phoned a couple of people and I went and saw somebody. But I wasn't very happy. Well, my son didn't look happy. You know, you have to go in there and know your child is content . . . I thought he won't be happy there. Then I heard about someone from a friend of my neighbor's. She was just perfect. Danny just went in there . . . he was so happy. Later she took Suzanne also and both were as happy as larks there. Since then she's moved away.

Valerie, in full-time employment, used the links she made during the time she was not employed and at home with her children to substitute care for each one in a different way. She attempted to maintain those aspects of her children's lives which she considered important for their happiness and social development, including friendships in the neighborhood and out-of-school activities. During the week she aimed to provide her children with experiences as close as possible to those they would have if she were at home. On the way to work, she took her eldest daughter to a friend in the neighborhood, from where the child was taken to school in a carpool with other local children. One day a week the child was brought back by carpool and spent some time at a friend's house before attending the local Brownie pack with her and other children from the area. After school on other days the child went to an after-school program. Valerie saw this as what she described as the one "weak link" in her arrangement, due to the social aspects of its location in a low-cost housing area, very different from her own:

I've found out through people I know and through the school, about what is called a Latch Key Program after school. It has its advantages and its drawbacks. The advantages are: it's structured time; they're not sitting in somebody's living room watching T.V. A good snack is provided and they do chores so they're learning. The bad thing is you have no control over the type of children attending. I try to explain to my daughter that a lot of the children there don't have the normal family background that she does.

Valerie also dropped her youngest daughter off at a licensed child care center on her way to work, from which the child is taken to preschool twice a week. Valerie had previously been a member of a local babysitting cooperative which was the source of information about the center, and her account of finding this expressed her concern for meeting the two children's differing needs:

A friend, the one who originally put me onto this babysitting co-op had already checked the field out and knew of an excellent lady who lived in the area where my older daughter goes to school. I was really concerned about (my youngest daughter), because they spend so much time away with the daycare person. But she's in one of the best daycares I've seen in my life.

In essence, this woman, by making substitutions in terms of place, time and personnel, has continued her mothering work in a different form while she is absent from the neighborhood for long periods of time. Yet, in order to achieve this, a variety of options had been incorporated in reallocating these aspects of her domestic labor to different spaces.

The next case illustrates another form of reallocation of mothering work, this time centering on adjustments of the home space itself. Like the other women of the study who have combined paid employment, mothering activities and housekeeping, Cara had experienced a period of considerable stress. She talked of this experience as the turning point in the organization of her life:

I don't want to be the superwoman who does everything and is the perfect wife. I've tried and ended up unconscious on the hospital floor. . . . My job's important to me, but so is my relationship with my husband and my children.

At the time of the interview, Cara had been a realtor for a year working long, but flexible hours. Although she had found it hard to make friends when she first lived in the area, by the time she took her current job, she had established firm social links with other women in the neighborhood. These links had been utilized on an irregular and not necessarily frequent basis, but they formed the fabric of her livelihood in both work in the formal economy and in child rearing. They provided sources of emergency babysitting, information on child care, and, in a few instances, had been sources of her sales. The family moved to a home with a den and extra bedroom so that Cara could work at home one day a week and employ a live-in help. Through a friend who ran a nanny agency, she hired a young woman from Europe to look after the children, and to provide substitutions for out-of-school activities, such as through playtimes and teaching the children French. On the nanny's days off, and when Cara was not at home, her husband looked after the children and helped with household tasks. Despite an extensive delegation of everyday tasks, Cara continued to see herself as the primary orchestrator and manager of the household. The adjustment of her children's activities and her domestic arrangements allowed her to be spatially separate from her children, while still ensuring their social development. Yet, even within this situation in which Cara appeared to have achieved a high degree of control of her working conditions, she experienced conflict. She commented: "It's something you never really work out, and I feel guilty sometimes, having someone else do the things with your kids that you'd be doing."

The women who were away from their children for considerable periods of time, adjusted the constraints of time and physical distance through the use of a variety of care resources which permitted them a flexible relationship to home and waged work environments. Although there were differences in socioeconomic standing which constrained the precise options exercised, in general the process of finding child care space involved tapping neighborhood resources. What was more difficult to harmonize was the moral dimension of the relationship of their child to neighborhood space. Had the women found a "safe" place for their children which would cater for not only their physical needs, but also provide an emotionally and socially nurturing environment? This was achieved in care situations which reflected dominant norms of nuclear family life and maintained a mother's values during her absence from home, and which had been found through trusted personal channels. A woman could continue to define herself as a good mother, providing for a variety of her children's needs, while at the same time reaching for the personal and economic rewards of the wage workplace. It was through the daily routines surrounding her child-centered activity that she was able to create support systems by which she could delegate part of her mothering duties to other people in other spaces.

CONCLUSION

This chapter has focused on the management strategies of women faced with the competing claims of wage labor and motherhood, which involve not only a matter of logistics as women juggle multiple roles, but also a normative dimension. For these women, the finding of quality, accessible child care involved less of a concern for location of settings and qualified personnel than that these care situations were spaces replicating, as near as possible, conditions of those at home. Licensing and paper credentials were of secondary importance to the use of the trusted channels of information, created in the context of the women's everyday work as mothers. Control over the conditions of the domestic workplace, in the sense of gaining access to space safe for their children, in both physical and social terms, was central to the ability of the women to engage in effective management strategies.

These strategies, however, are only one face of the coin, for social and economic conditions beyond the individual woman's control also shape the context in which informal solutions to child care are met. The need for flexible, low-cost care is predicated on the social conditions accompanying a gender division of labor which keeps women, in the main, as secondary workers who add wage work to the constancy of the job of mothering. The level of commitment to child care provision by the state reflects this aspect of women's relationship to the wage labor force, further delineating the conditions under which domestic and wage labor are carried out. At the same time, it is through women working together in the local conditions of a residential suburb, socially constructed in time and space in terms of a separation of different types of labor, that the development of women-based support resources is facilitated. Children do constrain the nature of women's participation in the wage labor force, but as Tietjen (1985) suggested, are also the avenues of support. It is through their gender-specific work that women are able to generate, and use support resources.

This study shows women actively confronting their problems, seeking to modify the conflicts engendered by the transitions from domestic work roles to those of wage worker within the unique social and geographic contexts of their lives. Flexibility is the key to how conflicting claims are met, but in general this is not found within the rubric of publicly provided services. In meeting their needs, the women use mechanisms which are culturally sanctioned according to local norms. The use of known caregivers working from family homes, enables the women to allay their concerns over remaining "good mothers," although the instability of care situations may continue to be a source of stress. Motherhood remains important to the women and is the central focus of many of their lives; but how this is interpreted and lived is forged within conditions emanating from the local and a larger gender division of labor within society, reflected in the occupational transformations accompanying shifting economic and social conditions. Women do not, however, experience these changes as passive receivers, but renegotiate their mothering work within local social networks.

While it is not possible to generalize the findings of this study cross-culturally, across classes or regionally, the centrality of the mother–child relationship to many women's lives and the commonality of their situation, as defined within a gender differentiation of labor, suggests that women's management strategies are likely to incorporate a moral as well as a practical dilemma. Local conditions will specify the framework within which particular strategies are developed, including the extent of the provision of public child care, but it would seem plausible to suggest that stability and ongoing care, such as with a single caregiver, is unlikely to be found where care providers are in a similar position to those for whom they are caring, that is, in transition between different types of labor. Moreover, given the overarching circumstances of a well-defined gender division of labor and the continuance of women taking the major responsibility for household activities, flexibility and continuity of care are necessary adjuncts to women's ability to accommodate the contingencies of their lives.

As increasing numbers of mothers work outside the home, knowledge of how a variety of care solutions are developed is important if we are to provide a stable form of care for children as their mothers work. Quality, access and affordability are essential aspects of this care, but the specific forms these take, and how these terms are defined need to be sensitive to local needs and environments. Active participation of women using such care would seem essential in adapting more global recommendations to fit the unique concerns of mothers living within specific geographical areas. It is these that contain the specific forms of work, family and neighborhood relations which shape both women's understanding of, and responses to their mothering work. Child care is a national issue, but its form must be as flexible and locally determined as the needs of the women it is created to serve. A diversity of care options appropriate to both children's and women's needs has the potential for restructuring women's geographies in a manner that can support a bridging of traditional spatial divisions of home and paid work.

Part V

CHILD CARE IN INTERNATIONAL PERSPECTIVE

10

THE STATE AND CHILD CARE
An international review from a geographical perspective[1]

Ruth Fincher

Policy and political discussion of governments' role in child care has probably never been greater in western democracies, yet geographers have made little contribution to the literature on the issue. At the same time the major spurt of geographic literature on public services provision (reviewed, for example, by Pinch, 1985) was marked by its lack of inclusion of child care among the public services to which access was calculated. This chapter seeks to review some of the contemporary English-speaking, social science literature on the state's intervention in child care, and also to isolate perspectives on the topic that feminist geographers could develop. Examples from a range of different countries will be provided to illustrate international diversities or uniformities in governments' involvement in child care provision.

It is difficult to isolate a specifically geographic standpoint on issues of child care. Shared geographic concerns on the matter have, however, been cultivated in feminist geographers' research on the close links between the extent of women's domestic labor, the particular forms of their paid employment, and their use or non-use of certain government- or non-government-provided "services" (including those that assist in the care of dependents) (see Fincher, 1989; Rose, 1990). A major interest of geographers then, has been to analyze the interdependencies of women's everyday activities in particular places. This interest has generally excluded extensive consideration of the characteristics of national policies about child care, except as a backdrop to what happens "on the ground" in the daily lives of people. Geographers considering "the state" in child care provision have, accordingly, focused on the implementation of government policies at the local level, and its distributional outcomes.

Looking more generally at literature on child care and the role of state intervention in its provision, there is usually a focus on national policies and on the national policy regimes that influence local outcomes rather than on those local outcomes themselves. "The state" in these accounts is not the site at which local negotiations occur over national child care provisions, but the national-level policy-setting process itself. I wish to organize a review of this broader literature in terms of the sorts of questions which feminist geographers might consider, even if the studies reviewed are not pitched at the scale at which most feminist geographers study child care. Four themes will be developed.

The first theme is the links of national child care policies to other policy priorities. It reflects (though at a different scale of enquiry) geographic interest in the interconnectedness of different activities of people's lives – the fact that workplace involvement does not occur in isolation from domestic commitments, and so on. Knowing how a child care policy is linked to other government platforms will suggest whether a government's child care policy is really a means to enhance the sort of labor market a government wishes (with trained women remaining in employment rather than leaving the labor force after they become mothers) or is a policy developed to encourage population growth or a certain "family" concept. In turn, this may suggest much about the way that child care will actually improve or limit women's (and men's) everyday life options. It will also signal which groups of men and women (and/or children) will benefit or lose out from child care interventions. In the first section, I take up this theme, discussing different "philosophies" of child care, by which I mean the ways that national governments justify their intervention in child care and their precise implementation strategies.

Second, feminist geographers have long acknowledged the need to consider the diversity of women, and less frequently men, who use child care services. "Women" are not homogeneous users of child care, nor do all "women" have a homogeneous interest in child care policies. A woman's class, ethnicity and residential location make a difference to her interest in, and capacity to use child care provisions. Accordingly, a concern with the varied accessibility of child care to different groups calls into question the adequacy of the representation of a monolithic "child care parent user," that is implicit in national policies. Who are the people that child care policies, implicitly or explicitly, cater for? In the second section, I consider the people represented in a range of child care policies, and the implicit definitions of "need" present in those policies.

Third, geographers have had a fluctuating interest in the state, its forms and functions. Usually, this has called for a concept of the state at the local level, at which scale general government policy is hard to separate from techniques of implementation and negotiation over policy outcomes in specific places. At this local level, the notion of the state, as a rigid set of internally consistent bureaucratic structures, has long been rejected by geographers. Instead, the state is viewed as a set of shifting, often contradictory government engagements with activists of various kinds, and even of conflicts between bureaucrats and elected officials at different levels of government (Chouinard and Fincher, 1987; Fincher, 1991). Geographic questioning of the state at the national level of policy formation, as at the local level, would seek out the conflicts that have given rise to particular policy stands. In particular, such questioning would try to determine the role in the policy-crafting process of citizens' (particularly women's) activism in favor of child care of certain forms. The third section takes up the issue of activism around child care in different places, trying to ascertain what forms of activism have succeeded where, or if the conflict forums of states in different countries have resulted in the appropriation of the activism of certain child care interest groups in ways those interest groups would not have wanted.

Finally, there has been effort by geographers to define the degree of government involvement in "public services." Typologies have been developed to characterize the alliance of governments with private sector, voluntary or charitable organizations, in the supply of what we often think of as "public services" (see Pinch, 1989; Rose, 1990; Wolch, 1990). This work is underpinned by an interest in the distributional implications of the forms of provision of services for diverse groups as well as an interest in reconceptualizing the familiar concepts of "public goods" and "collective consumption." Does it matter to different people if their "services" are provided directly by governments, or by private sector, for-profit agencies without government funding or regulation, or by any combination of these? In the final section, some of the actual mechanisms of child care provision and their distributional consequences are considered.

STATES' PHILOSOPHIES OF CHILD CARE PROVISION

A range of philosophies underpin different states' interventions in child care, affecting the degree and form of those interventions. These philosophies range from reasoning that government should or should not intervene in child care, to the specific designation of the policy context in which child care should be embedded. Of course, the real policy philosophy for child care may not be evident from public statements made justifying it – while politicians might laud child care provisions because they "liberate" women, for example, the better (if less politically beneficial) explanation might be that child care provisions retain expertise in the labor market at a time when labor supply is limited.

This section will discuss four ways in which the philosophies of government intervention in child care seem to vary in different countries. Once implemented as policies, these philosophies ensure that child care provisions suit certain service users rather than others, and define "need" in particular ways.

145

Leave child care provision to the "market"

There are a range of views about state intervention which are used to justify very meagre government involvement in child care provision. These views may be purely the product of economic ideology (that there is a market in child care services which should determine the "choices" available), or of slightly more mixed reasoning that also sees the care of children as a "private" responsibility, something that takes place beyond government's proper sphere of activity. One recent international review has suggested the US and Britain as extreme examples of this, since they had government philosophies in the 1980s and early 1990s that resisted any "intrusion" in child care for fear of reducing "efficiency" (Lamb and Sternberg, 1992: 9).

Accordingly, in Britain, most children under 3 years old are cared for by their parents or relatives; children in those local authority nurseries that do exist are less likely to have employed parents, and are more likely to come from economically poorer households (Moss, 1991; also see Pinch, 1987). A particular view of appropriate parenting appears recently to have become very significant in Britain. Moss (1991) notes the persistence and dominance of the notion that children should be in the full-time care of their mothers until at least the age of 3, or else their development will be harmed. This view that child care is a private matter became very solidly entrenched during the Thatcher administration (and continues under Major), bolstered by a restatement of values of individualism and the "naturalness" of women devoting themselves to domestic duties in the home (Pinch, 1987). Such views, in fact, sit well with economic policies that have supported the government's efforts not to intervene in employed peoples' parenting responsibilities. Even in the face of growing labor shortages in the 1980s and 1990s, the Conservative British government has chosen not to become involved in child care, but rather to leave any child care provisions to non-government sectors (Moss, 1991; Ochiltree, 1991).

In the US, the last decade's reduction of federal government spending on a range of social programs was accompanied, in the case of child care, by debate over who should actually provide a system of child care services (Klein, 1992). Various government reports had noted the need for child care because of the realities of women's labor force participation, and indeed in the late 1960s federal legislation had made the involvement of the central government in child care a possibility. However, disagreement between the various existing providers (including churches, groups public and private within the social welfare system, corporations, public schools) within the "patchwork" child care set-up of the US, each trying to enhance their own power over child care, has hampered government-controlled provision (Klein, 1992). As this debate continued, initiatives of the Reagan administration in the early 1980s shifted the nature of child care delivery, from that in the 1970s with more existing centers operating on a non-profit basis, to the present situation in which the for-profit sector dominates (Klein, 1992; also see Bloom and Steen, this volume).

Elsewhere, the debates over government involvement in child care focus more on the appropriateness of economic rationalist notions that "the market" in child care should prevail and dictate choices available. In Australia, the argument has been made that state subsidy of child care causes that care to be too uniformly of high quality, leaving no scope for the user who wants lower cost care of a different style. Regulation of child care by governments is criticized for the same reason – it allows centers little flexibility in selecting the mix of staff they wish to employ or deciding to grow above a certain size – and therefore provides less choice for consumers (Raneberg and Daubney, 1991). However, these arguments have not won out in the discussion over federal government involvement in Australian child care in the 1980s. Strong argument from children's services and early childhood development specialists (for example, National Association of Community Based Children's Services, 1989) for federally regulated and subsidized child care centers seems to have aligned through the 1980s with the views of influential bureaucrats in the Labour government's federal departments on such matters (see Franzway *et al.*, 1989). But the notions of "flexibility" that often accompany market-choice arguments for government abstention from child care provision are often seductive, pointing as they do to the rigidities of centrally funded and regulated child care. In this context, Rose (1990) discusses the emergence of recommendations for greater diversity and flexibility in child care, made by government child care committees in Canada. Without universal government child care provision, Canada, like Australia, has a situation in which the government services that are provided are inaccessible to many – for reasons of cost, location, hours of opening and cultural preference regarding the style of care. Arguments about the need for flexibility are, then, attractive to many who wish to see greater accessibility to child care. On the other hand, flexibility may be sought by others via a reduction of regulated government involvement in, and commitment to, child care. It is against this latter interpretation of the best route to improved child care provision that many activists in Canada and Australia rail. In Australia, such activists are particularly concerned that the federal government might move to subsidize commercial child care centers, following its recent move to allow users of commercial centers to receive fee subsidies (Brennan, 1992).

Child care is integral to labor market policies

Moving to a second perspective, it is clear that in a number of countries government activity in the provision of child care has been associated with labor market policies. In particular, child care appears to be a product of government efforts to retain women's labor force participation, although whether this is for the benefit of the economy or the women (or children) is not always clear. In the US, the emergence of for-profit, workplace-based, child care is a direct result of the last decade's rejection of central government child care provision in a labor market in which women are a growing proportion of employees. (Government provision of child care has continued in the US, however, for "welfare mothers" participating in work programs (Martinez, 1989).) Miller (1990: 94) notes that the economics of employer-sponsored child care, because it is not regulated by specific government policies, is "expressed in the language of employer investment and employee performance . . . the formal evaluation of such care looks more for efficient employee behaviors and increases in productivity than at quality of day care services." For-profit child care providers, from one-operator providers to large chains, expanded at a more rapid rate during the 1980s in the US than did non-profit providers. Klein (1992) accounts for the growth of chains, in particular, in terms of the economic advantages available to their operators because of the growth in the number of middle-class employees (with incomes falling between those eligible for federal subsidies and those who could afford very expensive private care) who would be their clients. Tax advantages also accrued to both the employers providing private sector child care, and the employees using it.

In the US in the 1980s and early 1990s, then, the growth of private child care occurred in a government child care provision vacuum, at a time when labor market trends made it profitable for such new child care services to be available for the middle classes. In other countries, direct government participation in child care provision is linked closely to, and enhances, economic and labor market activities by those governments. Sweden's government is always presented as a prime example of this. Over much of the post-Second World War period it has used a range of social policies and services to encourage both higher birth rates and women's labor force participation (Adams and Winston, 1980; Haas, 1992). During the 1970s and 1980s, efforts were made to encourage female employment by increasing public expenditure on the expansion of child care places. By 1987, 47 per cent of Swedish preschoolers were cared for in the public child care system, although the cost of, and access to child care varied locally (Lewis and Astrom, 1992). This has been part of a dramatic shift in the design of welfare entitlements since the 1970s, in which a range of government provisions associated with parenting (parental leave and parental insurance being those most commonly discussed) became available to women and men on the basis of their labor force participation rather than on the assumption that men would be employees and women carers (Lewis and Astrom, 1992).

In Australia, the situation differs yet again. Although government-subsidized and regulated child care is not explicitly part of a package of policies providing assistance to employed parents, since the late 1980s such child care is increasingly justified by government spokespeople as an economic or labor market intervention. Child care is part of a policy discourse about paid work. Its economic benefits for government have been emphasized in terms of how it allows parents to train or participate in the paid labor force, helps employers by keeping a pool of workers accessible to the labor market, and ensures that the country retains skilled workers in the labor force (Blewett, 1989). Consistent with this labor market link, when a National Children's Services Advisory Council was established in 1991, one of its major concerns was to assess the labor market issues of relevance to families with children, with a view to retaining affordable child care so that parents could remain in the labor force (*Minor Matters*, 1991). In fact, Bennett (1991) has argued that the Australian federal child care scheme functions as a form of "industrial infrastructure," whose priority of access guidelines ensure that workforce participants and those in training or seeking paid work are served best and first. The labor market justification for child care policy has enabled the Australian government to move towards encouraging the expansion of the number of employer-sponsored, workplace-based child care centers. Australian employers so far, however, seem less than keen on the broader issue of appreciating their workers' family responsibilities, which a major move to their supply of child care places or subsidies would seem to demand (Wolcott, 1991).

The labor market rationale of contemporary Australian government child care expenditure has occasioned criticism regarding the continual association of child care with women's labor force participation opportunities. While it is true that women do take primary responsibility for caring for children, this association reinforces the nurturing role of women and the gender-based division of labor which gives them that responsibility (Press, 1991). Furthermore, it is ironic that while women's labor force participation justifies expenditure on child care in Australia, governments actively maintain the poor working conditions of child care workers (almost exclusively women) in these government-facilitated schemes. Bennett (1991) shows how government cost-cutting priorities have been used to justify the continued low cost of child care workers. Expanded child care has been provided alongside the dismantling of financial and legislative protection for child care workers, a deskilling process. Furthermore, the government has continued to exploit workers in the Family Day Care sector (individual workers who look after a small group of others' children in their own homes) by denying them legal access to the basic protections given to employees, and defining them (though they are controlled by government stipulations) as independent contractors (Bennett, 1991).

Child care as part of parenting or family policies

A third government view of child care associates it with a particular family policy orientation. I have identified the reliance of recent British

149

governments on the notion that families should be able to cope individually with their child care needs, which, in turn, justifies the lack of state intervention in child care provision. In the US, continued argument about which providers should dominate child care provision has allowed an ideology of minimal government intervention to characterize the child care "industry," in a political culture long ill at ease, officially, with acceptance of the family as a site of government activity (see Adams and Winston, 1980: Ch. 5). Badelt (1991) has a useful list of the political issues around which parenting issues have been debated and developed in Austria in the last couple of decades. These issues are probably common to many countries in which democratic states intervene in child care, though, of course, their resolution differs. They include: the extent to which "traditional" families (married, heterosexual parents with children) are valued over other forms of family; support for women (primarily) to engage in either paid work or work in the home; equality for women and its implications for their role in families; support for the economic circumstances of families with young children; and using family policies as part of a broader population policy aimed at expanding population numbers.

An interesting difference is now emerging in the policy views of different "sorts" of families in a number of countries, and the way their child care needs are being defined and provided for. In the US, the families being supported by child care provisions (usually by private sector, often corporate providers) are primarily those of working parents, members of the broadly designated middle classes. This child care support, as noted in the previous section, rests on private sector recognition of who is in paid work and needs child care. It also rests on a government view of the family that has reached "no consensus that all or even most women *should* work; therefore there is no agreement that women should be supported in their work and thus on a family–labor market policy for the middle class" (Spakes, 1992: 56). On the other hand, there is an emerging view that women from poor families *should* work, and that child care should be available to them *only* if they enter the labor market or training (Spakes, 1992). Martinez (1989), for example, chronicles the welfare reform legislation of the mid-1980s in which child care provisions for welfare recipients accompanied "workfare" requirements that mothers of very young children undertake paid work in order to receive welfare benefits and child care services.

Even if the sorts of family policies now suggested in the US (like parental leave, flexible working hours, employer- and government-subsidized child care, income tax deductions for child care) were instituted by the government (and there is strong resistance given that they are inconsistent with the broader American philosophies of government intervention), such policies would support primarily the "revised middle class family model – the dual wage earner, two-parent family – and . . . the single, working mother and her family" (Spakes, 1991: 34). They fail to include other forms of families or households, which may actually be precluded from receiving subsidized care (Spakes, 1991). Sweeney (1989) develops a similar point about the class-based nature of government provision in Australia, taking a broad focus including the systems of child welfare and family support, as well as child day care. She argues that the welfare state does not overcome inequalities in its family policies regarding children, but rather segments differences between poor and middle-class families. It directs coercive policies to families in poverty, often removing children from families without resources, thus helping those families improve their economic situation, but also exposing the children to role models from other classes. Yet it helps middle-class families with their child rearing tasks through a range of non-stigmatized family support services, including daily child care.

Perhaps it is in those European countries that have developed a set of "parenting" policies including child care, that the circumstances of families have received most consistent scrutiny, or at least the most direct policy response. Sweden stands out as the country with the most developed parental leave legislation in Europe, the result of which is that "the vast majority of women claim virtually the whole amount of permitted parental leave at the 90 per cent replacement of income rate" (Lewis and Astrom, 1992: 70). Haas (1992) describes Sweden's relatively long-standing concern with equal parenthood, with calls for women and men to share equally the economic and child care responsibilities of families. The reforms in Sweden that have resulted from the concern with equal parenthood include parental leave (which became law in 1974), but also alterations of previously maternal care practices to include fathers (for example, helping at childbirth), and provisions for fathers as well as mothers to have paid days off from work to care for sick children, to have some paid leave from work to visit schools or child care centers, to reduce the hours of the workday for child care reasons (Haas, 1992). Sweden's radical attempt to achieve gender equity via family policy has, however, been criticized recently by authors evaluating its outcomes. Though Swedish social policy has established means by which fathers may attend to children as much as mothers, have fathers actually taken up the opportunity? Mothers in Sweden have entered the paid labor force like fathers, but has the situation shifted, other than this? Haas (1992; also see Spakes, 1991) finds that gender equality has not been achieved with the Swedish family policy of parental leave because: (1) the labor market remains sex segregated, with women more likely to be in part-time jobs and, therefore, with continuing responsibility for children in the home; and (2) only a minority of fathers eligible for parental leave actually take it (although the numbers are increasing).

Other authors go further, arguing that the Swedish social policies directed at shaping family behavior toward gender equality, actually have exaggerated gender divisions (Lewis and Astrom, 1992). The Swedish strategy has defined all adults as workers, supported women's entry into paid work, and then compensated both women and men for child-based absences from that work, thus "women were 'forced' to be workers by dint of both the effects of the tax changes and by the changed basis of women's entitlements to benefits, [but] men were not 'forced' to be carers" (Lewis and Astrom, 1992: 73). Though this seems a policy strategy based on equal treatment for men and women (which has, perhaps inadvertently, had the effect of reinforcing women's dual caring *and* working roles), in fact it is also a policy strategy that women have manipulated to affirm their difference from men. Once having taken a job, women have been able to claim leave because they are mothers, and to stay at home for long periods (Lewis and Astrom, 1992: 77). Another paradox identified by Lewis and Astrom (1992:77) is that in the late 1980s, government promises to increase parental leave entitlements were associated with failure to provide sufficient government child care so that women could choose *not* to take the maximum amount of parental leave.

Child care as a women's issue

How are the philosophies implemented by governments in their child care provisions situated with respect to the view that child care is part of women's caring role as mothers? Critics of the continued gender inequalities in Sweden, as noted above, find that despite wide-ranging social and family policies aimed at encouraging men's assumption of caring tasks, women retain domestic duties, including major responsibility for children. In both Sweden and Denmark:

> women have been integrated into a labor market which is structured by a male norm; working conditions, hours of work, and, to a certain degree, wages, have relied upon the principle that the "normal" worker has somebody else to take care of the house and care work within the family. . . . The state still maintains a hierarchical and gendered division of work and it has, by means of child care policies, institutionalized women's double shift.
>
> (Borchorst, 1990)

In North America, US family policy is criticized for being pro-natalist, for primarily viewing women as bearers and carers of children without really addressing how the caring will be accomplished (Spakes, 1991). Canadian child care discourse has been criticized for failing to emphasize the central societal role of women's caring, but rather focusing on controversies over appropriate forms of service provision and government involvement (Ferguson, 1991). The taken-for-granted nature of women's caring work is reflected in the low wages of workers in government child care schemes, as well as other situations. Ironically, in Canada, child care workers have sought to raise the status of their work (and therefore pay levels) through emphasizing its educational component and minimizing the significance of its caring component (Ferguson, 1991). Norwegian research finds that the welfare state depends on "mothers and minders, who supplement or make substitutes for the shortages of the state-sponsored services" (Leira, 1990: 155). As the Norwegian welfare state expanded its provisions, it was assumed, says Leira (1990), that women would go on caring: "motherhood" remained a necessary ingredient of women's entry into new labor market positions and government support of that.

It is clear that across the range of national policy child care settings much caring of young children is done outside the organized care provisions that are available, whether people are users of formal child care services or not. Much of that caring falls to women, and the amount of work it involves is invisible and underestimated. The sharing of caring work between men and women is often suggested as a policy goal. As the Swedish case shows, however, while policies might push for such an outcome, other social structures may resist this.

REPRESENTATIONS OF CHILD CARE USERS IN GOVERNMENT PROVISIONS – WHO IS CATERED FOR?

Basic to the requirement that child care policies represent the diversity of child care users is a certain flexibility in service provision and a recognition that child care users may have different preferences about the forms of child care, and different capacities to pay for, or organize their access to those forms of child care. Where child care provision is centrally controlled by governments, a recognition of difference is difficult to incorporate. Unfortunately, as noted in the first section, the particular rigidities of centrally provided care can lead to claims that government abstention from child care, or reduction of government intervention in child care, is the best way to recognize diversity. What, then, are the representations of child care users that are found in policies? Across the English-language literature about the state's role in child care, the commonest complaints about the way users are portrayed in child care policies concern the class and ethnic characterizations of child care "needs." In this section I will explore the characteristics of the users that child care provisions best serve, or fail to serve.

Assessments of child care policies or stances developed in a range of countries in the 1980s indicate that they favor "middle-class" households, those (presumably) in which heterosexual women and men are earners of moderate incomes. Spakes (1991), for example, has claimed that any expansion in US government child care provisions along the lines of the parental provisions of Sweden, would benefit only such middle-class families.

Of course, it makes a difference to their access to child care whether an employed parent is a part-time or full-time worker. Dex and Shaw (1986) comment on the ways that women of different ages have been incorporated into the US and British economies, and the implications of this for their access to the forms of child care on offer. In Australia, shift workers are a group often excluded from child care centers and even family day care arrangements, because their hours of paid work are outside the hours of operation of child care services. A recent study of health care workers (primarily nurses) in Queensland found that the only parents able to work shifts in the health industry were those with family and friends willing to provide informal child care (Gatfield and Griffin, 1990). This was not entirely the result of lack of federal government foresight about such workers. It was, however, associated with the very high costs of extended-hours child care staffed by workers with salary loadings for those hours (especially because usage of a center-type service might not be high), and also with lack of interest in the issue on the part of hospital managements.

The cost of government-regulated and organized care is (as with the shift workers) also an issue, though the way the cost falls most heavily on certain groups varies in different places. In Australia and Canada (Fincher, 1993; Rose, 1990) the cost of care in government centers for users requiring 8 am to 5:30 pm care is higher than many households can afford, even with government fee subsidy schemes targeted to those below a certain income ceiling. Elsewhere, the situation is different. In both Sweden and the Netherlands, for example, child care centers are very heavily government-subsidized to permit access for low-income groups, and yet educated families, who prefer center-based child care, make up a large proportion of the users of those centers (Clerkx and Van Ijzendoorn, 1992; Hwang and Broberg, 1992). In Sweden, this has caused some political opposition because people (for instance, those in areas without access to such centers) receive no subsidy for taking care of their children themselves. In the Netherlands those who prefer center-based care resent government attempts to restrict those centers to people of low income. Generalizations about the preferences of different class groups have been made, with one Swedish study finding that:

> Working class parents consider some form of home-based child care to be ideal more often than parents from higher social classes. . . . The working class parents' relative unwillingness to place their children in municipal child care has to do with an emphasis on the value of parenthood, a desire to be close to one's child and mould its development, and doubt about the quality of collective child care.
>
> (Hwang and Broberg, 1992)

The subtlety and complexity with which workplace and family policies intertwine to benefit certain class groups over others is expressed in many ways other than the direct financial cost of care. In the US, where child care is provided by some employers to their workers, Starrels (1992) notes that "lower status employees" in firms, especially women, are likely to have less access to on-site child care and to flexible schedules that make parenting and employment easier. Furthermore, tax deductions for child care expenses fail to benefit many working mothers, who do not pay enough taxes to claim such a deduction (Starrels, 1992). Making a similar point about Canada, Rose (1990) notes how the tax deductions available to upper- and middle-income parents in Québec are worth far less to those on lower-incomes. My own research in Melbourne, Australia, emphasizes the contribution that parents must make to the establishment and running of government child care centers – contributions not just in monetary fees, but also in time and skills like writing submissions to government, participating in parent management committees and so on (Fincher, 1991). The way that child care centers are established and run thus rewards the skills of those who are English-speaking, used to participating in public forums and writing reports, and who have the time to give to these tasks (because they are employed part-time or have flexible jobs that they can fit these tasks around). It excludes those whose English language skills are limited, who have not the time (not having a suitably flexible job), nor the resources to write reports and participate in public discussion. These are class issues.

These are also issues of ethnicity. In fact it is often difficult to tell if the usage of child care services is due to preferences that might be ascribed to ethnic affiliations, or to lack of resources. In many assessments of child care use, however, the point is made that immigrants and indigenous peoples, who are not members of the dominant ethnic group, are less likely to use or to prefer government-regulated and organized child care services (see Rose, 1991). This suggests that certain people's needs are better represented, or catered for than others.

In the US, many groups advocating on behalf of child development and child care for Black children have long pressed for community child care rather than child care provision through the public school system (Klein, 1992). The National Black Child Development Institute has lobbied for the need for parent involvement in child care (which is less likely, they argue, if public schools are involved), and for teaching methods appropriate to very young children. They have also highlighted "the schools" tendency to socialize children according to White middle-class standards (Klein, 1992). In Sweden in the late 1980s, children of immigrant parents were slightly under-represented in government child care centers, with some suggestion made that immigrants from southern Europe "are more 'home oriented' than their Swedish counterparts" (Hwang and Broberg, 1992: 47). (The authors depicted members of the Swedish working class in the same fashion.) The situation in the Netherlands suggests that language skills may be significant in influencing whether particular immigrants place their children in government child care facilities. Dutch-speaking mothers from Surinam usually place their children in centers, but Moroccan and Turkish children are more often placed with family members, others from the neighborhood or "ethnic guestmothers" (Clerkx and Van Ijzendoorn, 1992). The circumstances of recent immigrants in Québec are strained because they are unable to afford formal child care, yet lack the extended families and parent networks to organize child care for themselves (Rose, 1990, 1991).

Occasionally there are innovative schemes reported that actively enhance childrens' ethnic difference. The Te Kohanga Reo child care initiative in New Zealand is one example. A Maori response to White monoculturalism, launched in 1982, it was catering to more than 6,000 children in over 400 local centers within three years. The scheme emphasizes child care arrangements that are specific to each locality, and that "practise total immersion – virtually from birth . . . of the current generation of children in Maori language, culture and wairua (spirituality)" (Smith and Swain, 1988: 110).

The major insensitivity to difference in states' child care provisions of the 1980s is usually reported to be about class and ethnicity. Two other matters, however, are significant. One has to do with life course: government policies tend to assume that those caring for small children are young families without other caring responsibilities (such as ailing, elderly parents). Fernandez (1990) makes this point, in particular, about the assumptions of corporate child care programs in the US, though the same point could be made of other child care contexts. The second matter, mentioned from time to time and always if the assessment is written by a geographer, is that of location. Child care schemes designed, regulated or run by governments tend to favor urban over rural dwellers (and indeed those in certain urban or regional locations over others). The difficulties faced by low-income parents and parents requiring culturally different (to the monocultural norm) forms of child care, are exacerbated greatly if they live in an unfavored location. Ferguson, writing about Canada (1991: 99) has insisted that one of several principles of child care change should: "always be assessed in terms of their implications for women of all classes, races and geographical locations." She finds that White, urban, higher-income women have benefited from child care provisions in the 1980s, and that licensed child care centers are especially scarce in Canada's rural areas and Indian reserves.

In places where there is some local or regional discretion over the use of government funds (perhaps central government funds) to establish child care facilities, the urban–rural distinction is not so simply drawn. In Italy, for example, there are regional variations in levels of provision of child care. Government- and parent-managed community child care centers are far more common in the north central regions than in the southern and island regions (Corsaro and Emiliani, 1992). Local or regional governments act upon their particular views of appropriate forms of child care, and child care provision varies locally accordingly. My work on different suburbs within Melbourne has shown differences in the propensity of local governments to take up the opportunity for federal funding of local child care (Fincher, 1991). In England, Pinch (1987) has documented great variations in child care provision between local authority areas (and also within them) such that those areas controlled by Labour councils have more state-funded child care centers and preschool classes, whereas the counties dominated by Conservative councils tend to provide child care through non-government agencies.

Spatial variation in child care opportunities can also, of course, be the direct product of factors other than government child care policies. Rose (1990) illustrates this, noting the tendency of immigrants in Montréal to live in certain suburbs without long-standing child care networks. This characteristic of the suburbs in question can be associated with gentrification and racial discrimination in housing. In Québec, there is no clear aim in government policy to distribute government-sponsored, non-profit child care centers in locations of particular need, except in rural areas (Rose, 1990). In Australia, in contrast, the National Children's Services Advisory Council aims to be more responsive to the needs of "different regions and communities. The aim will be to ensure services reach the areas where they are needed most" (*Minor Matters*, 1991: 7).

CHILD CARE ACTIVISM AND THE STATE

The long-standing interest of some geographers in state policies and institutions and how they are formed and changed, points to the question of how child care policies developed in the 1980s, given the varied philosophical underpinnings they exhibit. In particular, has women's activism in the state been responsible for the flood of political interest and government expenditure (in some countries) on expanding child care places? The hypothesis that the activism of women has played some part here seems plausible because the second wave of the women's movement in western countries preceded and accompanied changes in child care provisions. Certainly, research has indicated that in Britain, women's activism has been significant in advancing policies reflecting a broad range of women's interests in particular local regions (Mark-Lawson *et al.*, 1985). On the other hand, since we know that in many countries child care provision is associated with labor market policies and continuing pro-natalism, perhaps the emphasis on women's activism is misplaced.

What is the international evidence about the role of women's activism in prompting an increase in child care provision through the state? There are, not surprisingly, different views on this question, as well as different experiences in different places. I list four points of view expressed in the literature on this issue.

First, there are a number of claims that women's activism has been, in a general sense, important to the recent growth of state involvement in child care, even if it did not trigger the specific implementation of child care policies. In the Netherlands, feminist action groups are recorded as requesting government-backed children's creches in the early 1970s, and being resisted by a range of groups, child development specialists included (Clerkx and Van Ijzendoorn, 1992). In the 1980s, the Dutch women's movement re-entered the debate on the issue, despite the criticism of radical feminists who argued that child care facilities would limit men's responsibilities for sharing care of their children. The demands by the women's movement for child care occurred alongside a host of other social changes, including increases in divorce and remarriage and more positive attitudes toward child care centers (Clerkx and Van Ijzendoorn, 1992).

In her study of Québec, Rose (1990) notes the significance of the growing feminist movement in contributing to the political climate in which the Parti Québécois was elected in 1976. A concern with child care was part of the Parti Québécois's platform, thanks to the demands of feminists. The situation in Norway, where the strong representation of women in politics has not been accompanied by a rapid implementation of extensive child care policies, is examined by Kissman (1991). She claims, however, that women politicians have made a contribution in "changing the philosophy and values on which policies are developed, in the face of a relatively conservative political ideology in Norway as compared to Sweden and Denmark" (Kissman, 1991: 197).

Comparing China, the US and Sweden, Adams and Winston (1980) make the important point that there are different political traditions to consider. They argue that the model of women's activism as mass-based women's organizations lobbying receptive politicians may not apply in those places in which change must be accomplished from within governmental organizations and bureaucracies. Having made this careful qualification, they too find that women's movements in Sweden and China have not been responsible for the establishment of women's programs (including child care), but have been part of a larger process of growing interaction between policy makers and their constituents (Adams and Winston, 1980).

Second, it has been argued that women's activism has had no bearing whatever on the state's attention to child care and parental policies. Rather, the economic policies of national governments have dictated child care outcomes. The history of Sweden's family–labor market policies, argues Spakes (1992), indicates that the coalition of interest groups that developed policies to encourage men to take some family responsibilities and to allow women into the world of paid work by moving some of their caring work onto the state, did not include women activists.

Male-controlled and dominated labor unions played the major role in determining how to change men's attitudes toward family responsibilities and women's attitudes toward work. . . . thus controlling the dialogue, determining the agenda, and setting its pace.

(Spakes, 1992: 47)

Haas (1992) has a slightly different interpretation of the Swedish situation, however, indicating that by the time parental leave was legislated in Sweden (in 1974) women made up almost 25 per cent of the members of the Swedish parliament. This may have influenced the development of the parental leave policy. Furthermore, she states that the women currently in that parliament (33 per cent of its members) actively advocate women's and men's equal participation in child care.

Spakes (1992) also argues that the feminist movement in the US was not responsible for the increased entry of American women into the paid labor force. Like Swedish women, American women are largely employed in part-time work, in segments of the labor force in which women predominate, and in low-paid occupations. These are clearly things that no feminist movement would have agitated for.

Third, where women's activism for child care has occurred primarily within the state apparatus, there is mixed opinion as to the extent of the gains made. This is because in some cases in which feminists have been active within the state to expand child care provision, the original aims of that activism may have been reinterpreted to fit the state's different priorities or subsumed under more general policy goals. In Sweden, claims Haas (1992), feminists have pursued legislative politics as the road to improving women's circumstances and have emphasized family and labor market policies. The provision of child care, specifically, has been secondary to these overall policy goals under which women continue to do more domestic labor and to take more of the parental leaves. With little tradition in Sweden of a women's movement outside the state, any emerging critique of this situation seems to lack potential.

In Australia, much recent activism for the extension of child care has occurred within the state, though without Sweden's explicit tradition of gender equality. And in the 1980s, like Sweden in the past, the Australian federal government joined with the union movement (the Australian Council of Trade Unions) to redefine the social wage and the way it would change. Child care was included explicitly in the social wage over this period, but privatized child care was largely rejected (Gifford, 1992). Since the 1970s, "femocrats" (feminists employed in influential policy positions within the Australian state) have struggled, with considerable success, to effect state provision of child care in the form they have preferred (Franzway et al., 1989). Though a significant community-based child care lobby remains, its work seems always to be the countering of government attempts to enable the expansion of private, for-profit, child care and expand the cheapest forms of child care (those family day care schemes relying on poorly paid female labor), and to be defending those parts of the government program with which it agrees (for example, accreditation of centers which are to receive any government subsidy) (see for example Brennan, 1989; Ruchel, 1990). This means that the debate on child care and how it can be improved is primarily conducted in terms of the federal government's involvement and directions, rather than, for example, about whether men share sufficiently in child care with women. Still, the influence of femocrats and feminist academics in shaping Australia's recent social policy has been very considerable. In 1993, rather than moving to tax deductibility for child care expenses, a policy which (as already noted) benefits the middle class and has long been opposed by child care activists, strategies were adopted that allowed for cash rebates for child care expenses.

Canadian analysts of the influence of the state on feminist lobbying have pointed out the consequent control by the state over the language of the debate, and the terms in which it can be discussed (see Barnsley, 1988). Ferguson (1991) has complained of the limitations of discussion of child care there. She argues that attention has centered on forms of provision and the extent of government intervention rather than on who does the central work of caring in all these situations (an issue of less significance if a debate concerns only the mechanics of child care delivery by non-family members). Indeed Prentice (1988) points out that:

> In a span of little over a decade, childcare has moved from being a radical and militant demand of feminists and socialists to a legitimate issue on the social policy agenda of the state. . . . Yet a number of the [state's] initiatives contradict the movement's original demands for state support for childcare. . . . Although there are still many different voices within the childcare movement, the critical perspective which demanded transformed relations through the socialization of childcare is now barely a whisper.
>
> (Prentice, 1988: 59)

There seems evidence from Canada, then, as well as from Sweden, that the "mainstreaming" of child care (as Prentice, 1988, terms it) within the state is associated with failure to tackle a real issue underpinning the widespread demand for child care. The issue is that women continue to undertake much of the caring work, and that this caring work remains "women's work," largely not shared by men, despite the tremendous growth in women's paid employment.

Certainly, in Australia too, the state's dominance can be difficult to negotiate for feminist activists still interested in improving child care arrangements. This is especially the case when any criticism of the current, relatively generous (by the standards of the past) government provisions are likely to be used as political ammunition by those who wish for reductions in government involvement and expenditure, and reductions in the overall provision of child care.

Fourth, feminists have long accepted the view that political activism occurs in communities and homes as well as within the state and the formal political structures of our societies (see, for example, Bookman and Morgen, 1988). It should be emphasized that women's activism in child care, even in those countries in which the state has an important role in child care provision and regulation, is not just political activity within state bureaucracies or parliaments. (There is, perhaps, the danger of taking a restricted view of political activism, when focusing on the state and child care in the manner of this chapter.) A broad range in what constitutes political activism is nowhere clearer than in the US, where state involvement in child care remains limited and contested. There, it seems that child care *activism* at the local level often takes the form of direct service *provision*, of the establishment of alternative, community-owned, child care services. This is service provision as well as activism, for provision of community-owned services is accompanied by advocacy of certain rights for children, women and families. As Kahn and Kamerman (1987b) note:

> Faced by growing needs for child care for the young children of working parents, at a time of federal cutbacks and changed regulations, many communities have attempted to package new solutions. And, while focused on doing what seemed necessary and possible, they have also evolved new local structures for child care operations and leadership.
>
> (1987b: 30)

Local women's rights policy networks have emerged within some US cities, which have established services for women and children, including child care, that are consistent with a broader feminist-inspired family policy agenda (Boles, 1989). Many municipal child care centers, often run as parents' cooperatives, have been established by local women's rights activists. But, notes Boles (1989) local governments have not been threatened by these "alternative" services, since they have no desire to dominate this costly area of service provision.

FORMS OF GOVERNMENT CHILD CARE PROVISION AND THEIR DISTRIBUTIONAL IMPLICATIONS

In the second section, I described the sorts of child care users for whom different governments' child care philosophies and policies are designed. Government child care interventions, in general, seem to assist middle-class employees who speak the dominant language and are members of the dominant ethnic group in the country in question.

In this section, however, I move on to ask what difference it makes if these services are wholly provided by government-organized and subsidized bodies, by providers in which the public and community sectors are in partnership, or by organizations run for profit by a range of commercial enterprises? Much discussion turns on this question. I have argued also that some feminists are critical of the silence surrounding the assumption that women will continue to have primary responsibility for child care, a matter which should be linked explicitly to discussions of the form of provision of that care. It is beyond the scope of this chapter to consider the variety of arrangements made by people to care for their children, but it is clear that regardless of the provisions of the state, the majority of young children everywhere are cared for informally, and by women.

There are a number of international overviews of the details of different countries' child care provisions (see, for example Kamerman, 1989, 1991). Such accounts discuss the varied policies at a level of detail beyond my brief in this chapter. What I will do, however, is raise three issues over which debate continues about the difference that varied interventions in child care make. As I describe these, it should be stressed that the success or failure of different child care policy mixes will vary with the national policy culture and the broader national socioeconomic context in which they are embedded. It will be significant for the effects of child care interventions on reducing women's caring responsibilities, for example, if there are labor market conditions, like persistent low wages for women, that make those women unable to afford adequate child care.

For-profit provision of child care

First, I note the issue of for-profit provision of child care. This can occur with or without government intervention: in the US, for-profit child care seems, more than elsewhere, to be outside government regulation or control (though it has been fostered by the failure of government to establish a prominent government role in child care, see Tuominen, 1991). Elsewhere, in Australia and Canada for example, for-profit child care is subject to government regulation.

For-profit provision without government oversight could be said to occur in separate child care enterprises (the "child care industry") and also within workplace-based child care situations. In the US, so-called proprietary providers in the child care industry are themselves diverse, including church-sponsored enterprises and centers that are part of major commercial child care chains rather than independent operators' establishments (Klein, 1992). The central question about the relative merits of non-profit and for-profit child care in the US (as elsewhere) has been whether high quality child care can also be profit-making child care. Klein (1992) suggests that there are two major criticisms of for-profit care: (1) quality of care is limited because for-profit child care enterprises save money by underpaying staff or limiting their numbers or training; and (2) the large group sizes and poor staff-to-children ratios sometimes associated with for-profit child care establishments are detrimental to children's development.

Research in the US about this issue has concluded that for-profit programs are not necessarily of poor quality, but that there is reason for concern about the possibility of poor service in for-profit firms (Klein, 1992). Others concur, finding "for-profit status is a main predictor of poorer quality" in child care (quoted in Tuominen, 1991: 463). An indication that the answer to this question turns on the quality of staff employed in the centers comes from American research on child care center workers. This research, though not just about workers in for-profit centers, shows that low staff wages are associated with high staff turnover and with low quality of care for children (Phillips *et al.*, 1991). If it is primarily "middle-class" families using these child care centers, with poorer families using those government institutions provided for clients of the welfare state, then lower-income members of this middle class will presumably be obtaining poorer quality child care in the for-profit sector, governed by the "you pay for what you get" reality. At the same time, the US's social welfare system has been dismantled in the last decade, and welfare benefits previously available to poor women and their children have been reduced (Tuominen, 1991). It has been claimed, therefore, that privatization and the decline of the welfare state have reduced child care provisions for lower-income people, and expanded them for higher-income people.

One means by which child care for the better-off has been expanded in the US has been the growth of workplace child care, not "for-profit" in and of itself, but certainly "for-profit" in enhancing the productivity of those firms' employees. I have already indicated how low-status women employees are less likely to have access to the flexible family policies and in-house child care programs of corporations. Starrels (1992) notes also that women employees are likely to be the major users of any such "family sensitive" workplace policies. Indeed, male employees and supervisors often exert subtle pressure for family matters to remain women's priority, thus subverting the apparent gender neutrality of these provisions. The access of many parents to such workplace provisions, especially lower-income women and, of course, those who do not work within a firm with such family policies, is minimal.

In Australia, these sorts of findings about the distributional impacts of increased privatization of child care in the US are used by activists to try and forestall growing government acceptance of a for-profit child care sector. Ruchel (1990), for example, describes the opposition of the community-based child care sector to the federal government's decision to extend fee subsidies to users of commercial child care centers, and its fear that the federal government no longer sees a significant difference between non-profit (publicly-funded) and for-profit services. The decision to allow fee subsidies to users of commercial centers must be seen in part as a government response to the demands of the for-profit child care industry, and the more recent decision to require accreditation for such centers as a response to the continuing demands of the lobby that supports government provision of non-profit care. Perhaps privatized child care will be less associated with poor quality if it meets the standards set by government. In this event, government subsidy (through fee subsidy) of commercial child care operations may ensure better distributional effects than if government subsidies were reserved for non-profit child care, leaving the commercial sector to regulate itself.

Regulation of child care

The second issue to be debated is whether governments should regulate child care, even if it is not subsidizing it. The strongest voice arguing for regulation of child care comes from those concerned with child development. They argue that children's development is best fostered in certain physical conditions, in limiting the size of the groups of children and so on. Generally, regulations are discussed in the context of center care, rather than family day care situations. Regulations are often opposed by "free market" advocates, who insist that they increase the costs of child care and restrict all parent users to the same high standards (this argument was described in the first section). Nelson (1991), however, provides interesting insights into the views of some American family day care providers on the issue of their regulation. They understand that regulation of family day care and the training of its providers is advocated as a means to protect consumers, by ensuring certain standards of safety in homes, and limiting the number of children looked after in any one home. These family day caregivers object to certain aspects of being regulated, however. They feel that regulation will make their relationships with children less flexible and more formal. They may resist the requirement that they undergo training, seeing this as rejection of skills they have already acquired over many years on the job as family child caregivers. And they want to avoid regulation changing the nature of the caring environment they provide, which they feel has warmth and intimacy, and does not "push" children the way that more formal center-based child care arrangements do (Nelson, 1991). Other home-based caregivers see registration and regulation as a resource, allowing them access to lending libraries of books and toys, for example.

The message is clear that central government regulation is not instantly of benefit to all child care providers, and it generates resistance from those who have previously been unregulated. It is interesting that central regulation of child care is also criticized (for different reasons) by advocates of government intervention in child care, and by those who consider universal, non-profit child care their goal. Accommodating the demands of all these groups is a vexing political task.

The relationship between formal and informal services

In discussing what difference it makes how child care is provided, there is one important issue on which there has been little research: the relationship between formal and informal services. It is often assumed that the existence of a large formal sector in social services provision will discourage the growth of alternative, informal means of service provision (d'Abbs, 1991). Such an expectation certainly seems to underpin the "free market" school of thought about child care – that if the government provides too much child care and regulates it too much, then no other forms of care will be able to appear. In contrast, d'Abbs (1991) suggests that if there is a formal sector providing organized and regulated services (though he does not speak specifically of child care) it is likely to encourage informal support. That is "mutual aid through informal networks is likely to be more extensive and more effective if it is backed up by adequate formal services" (d'Abbs, 1991: 128). It is quite conceivable, thinking along these lines, that the sharing of child care tasks between women and men within families may be encouraged if they are able to use appropriate formal child care services, than if they are left to their own devices to organize that care. Rose (1990, 1991) has hinted at the need for further enquiry into the relationship between formal and informal child care provisions in her Canadian study. She claims that a lack of formal, appropriate services means that women will end up continuing to do more of the caring. Her particular concern is that when the extended immigrant family is relied upon to substitute for government-funded child care services, this reinforces as "natural" the view that women are the only appropriate carers (Rose, 1990). The relationship between formal child care provision and the growth of informal support services in child care is a vital area for further research (also see Dyck, this volume, and Mackenzie and Truelove, 1993).

CONCLUSION

In this review chapter, I have identified several issues that might shed light on the current social science debates regarding the role of national governments in child care provision. Many of the issues discussed parallel recent concerns of feminist geographers, who have studied the situations of child care users primarily in local contexts, making a focus on the social relations of localities. Thus the strength that geographers bring to these debates is a deeper understanding of the local embeddedness of child care use and provision, within the context of a range of other local policy practices and daily life activities. It is this situation of child care in other social relations – an understanding of which is thoroughly present in geographic research – that leads me to conclude with the following two points about further research.

First, the geographic comprehension of the range of situations of which child care is a part, allows valuable broad perspectives on child care to be raised. It is important to keep placing our analyses of child care within wider questions like the distribution of caring responsibilities within the gender division of labor. A concern with forms of child care provision and their distributional consequences must not be permitted to mask this central issue about child care (and indeed all forms of caring), which is that it must be spread between men and women more evenly. Elsewhere (Fincher, 1993) I have tried to counterpose the circumstances of employed women who have dependents of different characteristics (young and aged); merely posing a question which talks about young and aged dependents in the one breath is outside the realm of those who study "child care" or "aged care" as separate social policy discourses. Thinking geographically helps direct the formation of these broader questions about the phenomenon of child care, limiting the tendency of child care discussion to become too focused on the operational details of specific policies and too separated from the bigger issues of caring work and its societal distribution. It is important that the imaginative centering of child care issues in a range of social justice and gender equity discourses should continue.

Second, the matter raised in the final Section, of the relationship between formal and informal child care provision in different policy and spatial contexts, needs further exploration. It is important to establish, with proper subtlety and empirical detail, the way that government-regulated, formally provided child care is associated with better informal systems that depend less on the labors of women. Also, the reverse situation needs exposure. We need to know if it is indeed true (with study of a range of different circumstances), as has been hypothesized, that women end up doing more caring work if they are unable to access suitable child care alternatives in the formal sector. It is a feminist priority to know whether and how the provisions of national states can be beneficial to a diverse range of users of child care. Similarly, more documentation is needed to show whether or how the conservative alternative of withdrawal (or minimization) of government support of child care may be worse for users of child care (women in particular) than the centralized, somewhat rigid, forms of state intervention in child care that presently exist.

Part VI

CONCLUSION

11

CONCLUSION

Kim England

While editing this book, many people have asked me how child care is geographical. As Marie Truelove (Chapter 7) argues, "(g)eographical analysis can contribute to an understanding of the effects of present and proposed policies on child care, particularly their spatial and redistributive impacts" (p. 106). Ruth Fincher suggested that "the strength that geographers bring to (debates regarding the role of national governments on child care provision) is a deeper understanding of the local embeddedness of child care use and provision, within the context of a range of other local policy practices and daily life activities" (p. 166). And Isabel Dyck, drawing on interviews with suburban women, remarked that in their attempts to resolve the dilemmas created by combining mothering with wage labor "the solutions they reached, in the form of 'safe space' for their children, were rooted in their work in the domestic workplaces of home and neighborhood" (p. 124). In bringing together the chapters in this book I sought to provide explorations of the geographies of child care and working mothers at different scales within the Canadian and US contexts, where, unlike most European Union and Scandinavian countries, there are no national child care or family policies. The US and Canada's patchwork provision mean that there are noticeable variations in child care; variations that are not related to local responses to local variations in child care need. As the contributors to this book showed, these variations have sociospatial implications: for instance, there are clear spatial differences in terms of delivery, and the type of child care that a child experiences is greatly dependent on where they live and their family's social characteristics.

In this chapter I outline a number of themes that were raised by the authors. I develop four themes. The first theme focuses on human agency and the activism of women and mothers. Second, I look at the gendered nature of child care. The third theme explores the ways that diversity intersects with child care. The final theme examines the different manifestations of geographies and scale that were adopted in the book.

WOMEN'S ACTIVISM AND HUMAN AGENCY

One thread running through a number of the chapters was the importance of women's political activism and their involvement in initiating child care, as well as the centrality of their agency in constructing their everyday geographies. Historically, as Truelove (Chapter 3) indicated, non-parental child care has been initiated by women. Ian Skelton's analysis of contemporary child care provision in Ontario indicated that women are much more likely than men to push for more child care services and that "the ratio of women to men on the boards of non-profit child care centers was recently estimated to be seven to two" (p. 65).

Fincher argued that child care is now an important topic of government regulation and election promises, such that child care has become "mainstreamed" as a political issue (also see Prentice, 1988; Rose, 1990). However, she also carefully considered the role of women's activism in increasing state interventions in child care provision. She found that there is evidence for and against the claim that women's activism triggered the recent growth of government involvement in child care. Moreover, other evidence suggests that some of the activists' goals have been appropriated or implemented by the state in ways unintended by the activists. Of course, activism does not only occur within formal political structures, but also informally in communities and homes, especially in terms of organizing and providing alternative, community-based child care in the context of limited government provision. Although Dyck dealt with women's local support networks in providing informal care rather than formal community-based care, she emphasized "women's use of private solutions to gaps in public provision" (p. 124).

By highlighting women's solutions, Dyck joins other authors in this volume (notably England, Myers-Jones and Brooker-Gross) who examined the complex juggling associated with combining mothering with paid employment. These analyses emphasized the ability of women to create strategies that enable them to negotiate, and even alter, the web of sociospatial relations and structures that shape their lives. Some of the women whom England interviewed constructed solutions that challenge conventional gender inequalities in the gender division of domestic labor. They had negotiated a more equitable division with partners and children, although many felt that child rearing and domestic responsibilities were ultimately their responsibility. These chapters reflect a persistent theme in feminist geographies – that women not be treated as passive receivers of wider socioeconomic relations, but that they be cast as human agents with transformative capacities within the context of broader sets of relations that structure people's lives.

THE GENDERING OF CHILD CARE

A number of the contributors pointed out that in the last twenty-five years the availability of child care rapidly increased and that child care has become "mainstreamed" as a political issue. However, Fincher pointed out that the underlying assumption of the nurturing role of women (as mothers or carers) has not been adequately challenged, and she questioned whether there has been sufficient erosion of the gendered assumptions that surround child care responsibilities. Even in Sweden, she noted, government policies assume that women are and will continue to be the major carers of children. And Skelton found that in Ontario women are much more likely than men to push for more child care services and to be involved in the management of child care centers their children attend.

Many of the authors drew on, and added to the plethora of studies that find that even when women are in full-time employment, child care (and housework) tend to remain the responsibility of women, including making child care arrangements, chauffeuring them to and from child care, and taking time off work when they are sick (see, for example: Hochschild with Machung, 1989; McDaniel, 1993). For instance, the local social networks that the women Dyck interviewed drew on for support, advice and informal child care were comprised entirely of other mothers. England, and Myers-Jones and Brooker-Gross argued that while rhetoric suggests that child care is increasingly identified as a "family issue" rather than a "women's issue," in practice child care remains women's work. This is the case whether women are full-time mothers or whether mothers (rather than fathers) modify their career aspirations or hours they work to accommodate their domestic responsibilities. Indeed, child care is often interpreted as enabling the woman in a heterosexual couple to engage in paid employment, a conclusion also reached by Gregson and Lowe (1994) in their study of the employment of nannies and cleaners by dual career, middle-class couples.

Generally, child rearing is constructed as mothering (rather than parenting or fathering) and child care continues to be identified as the responsibility of the mother and often gets cast in terms of mother substitution. Fincher showed how gendered assumptions and responsibilities are built into, and perpetuated by the form of government policies. However, as Dyck argued, cultural notions about motherhood and the centrality of mothering to certain constructions of femininity also percolate through informal care. She notes that "women face the paradox of a social contradiction in beliefs which simultaneously emphasize the importance of women caring for their young children at home, yet ascribe social status through success in the non-domestic world of paid work" (p. 124). Often then, the "best" child care is seen as that which most closely resembles that of a full-time mother and replicates conditions at home (Gregson and Lowe,

1994; Richardson, 1993). The potency of these ideologies helps explain the moral burden or sense of guilt felt by many of the working mothers interviewed by Dyck and England. The women interviewed by England often commented that they felt guilty: Joan said "I felt guilty all the time. I missed not being here when they got home from school, that made me feel guilty. I'm their mother, I'm supposed to be here" (p. 118). Dyck emphasized that the women she interviewed were anxious to remain good mothers while earning a wage and so developed child care arrangements that were "appropriate to their own interpretations of good mothering, and as close as possible to home conditions" (p. 133).

The fact that child care and child rearing remains "women's work" spills over into non-parental child care arrangements. For instance, Bloom and Steen, and Truelove (Chapter 3) pointed out that if relative care is adopted it is usually a grand*mother* who is the carer. However, the increased paid employment of women of all ages means that fewer grandmothers are available for informal care (also see Bowlby, 1990). This has greater impacts on some families than others as lone mothers and certain racialized groups are more dependent on relative care. At the same time, practically all non-relative child caregivers are women (over 95 per cent in Canada and the US, a pattern that is also typical of other countries, as Fincher showed). In the US at least, child care is also racialized: Bloom and Steen noted that Blacks and Hispanics are over-represented among child care workers relative to the total population. In addition, there has been an increase in foreign domestic workers and child care workers over the last ten years entering the US and Canada under special immigration programs (Bakan and Stasiulis, 1995; Glenn, 1992). The feminized, racialized and immigration aspects of child care work provide some explanation for the low wages that Bloom and Steen, and Fincher refer to. As they pointed out, child care providers are so poorly paid that their average salary falls below the poverty line. They often receive limited benefits and have poor working conditions. Not surprisingly turnover is high. Indeed, child care workers have the highest turnover rate of any occupation in the US (also see Ferguson, 1991; Melhuish and Moss, 1991b; O'Connell and Bloom, 1987).

DIVERSITIES AND CHILD CARE

Diversity was an important issue in terms of the differences among the women, families and children using child care. As Fincher remarked "'women' are not homogeneous users of child care, nor do all 'women' have a homogeneous interest in child care policies. A woman's class, ethnicity and residential location make a difference to her interest in, and capacity to use child care provisions" (p. 144). Previous research (Bean and Tienda, 1987; Farley and Allen, 1987; D. Rose, 1993) indicates that child care arrangements vary by "race" and ethnicity. But, as Fincher commented, it is difficult to ascertain whether this has to do with "cultural preferences" or limited subsidies and incomes. However, she also noted that there is recognition that there needs to be greater sensitivity to cultural differences in formal child care provision.

A number of the authors focused on differences by class or socio-economic status, noting that formal child care tends to privilege middle-class families. For instance, Skelton argued that class differences (and residential location) greatly influenced child care services in Ontario, and his analyses indicated that areas of high occupational status enjoyed higher levels of child care service. Myers-Jones and Brooker-Gross found that Blacksburg's three largest child care centers tended to serve upper middle-class families, few of the children at the centers were subsidized, and lower-income families were effectively priced out. Bloom and Steen, Fincher, and Truelove pointed out that changes to the Canadian and the US tax systems making child care costs deductible, advantaged middle-class families over lower-income families. Moreover, during the 1980s the US curtailed government involvement and encouraged the privatization of child care, further reinforcing middle-class advantage (however, as Bloom and Steen indicated, the 1990 Omnibus Budget Reconciliation Act contained a child care package intended to help low-income Americans).

That Canada and the US have relatively limited state involvement in child care means that parents often develop community-based child care arrangements. Fincher remarked that this practice is often linked with activism around certain rights for women, children and families. And Skelton highlighted the importance of parents' involvement for implement-ing and managing child care. However, as both authors point out, these activities are class issues. Skelton noted that the voluntary sector and the capacity to mount services are closely associated with high socioeconomic status. His analyses indicated that service levels in Ontario counties were much more closely associated with socioeconomic status than levels of need for child care (in terms of the labor force participation of lone parents and women with young children). Fincher noted that certain groups of parents are more able to contribute to establishing and running child care centers (English-speaking people with experience of public forums and writing reports) and are more able to devote time to these activities – and these groups often end up reaping greater rewards.

The increased use of formal care by middle-class parents has ignited a debate regarding the educational component of child care. A number of the authors allude to this (Fincher, Myers-Jones and Brooker-Gross, and Truelove). For example, over half of the respondents to Myers-Jones and Brooker-Gross's questionnaire held professional jobs, and 70 per cent of the parents responded that the educational value of the center was a reason that they selected the center. That early childhood education may be significant in a child's later development was central to Cromley's chapter. However, she directs our attention to an often overlooked aspect of this issue. She argued that the resurgence of interest in full-day kindergarten is not just related to the importance that middle-class parents attribute to their children's education and socialization, but also to the success of Project Head Start for improving basic skills of children from low-income families. She found that communities most likely to adopt full-day kindergarten were in urban areas with higher-income and middle-class residents (or rural areas where low densities and the cost of busing children make full-day kindergarten more economically sound than other options). While Cromley did not include a "race" variable, she points out that the state's African-American population is highly urbanized and, as such, are likely to be living in school districts offering full-day kindergarten.

GEOGRAPHIES, SCALE AND CHILD CARE

The contributors dealt with child care at a variety of scales. Bloom and Steen, Fincher, and Truelove (Chapter 3) examined child care at the national and regional level in their analyses of government involvement in child care (although Fincher and Truelove also looked at other scales as well). Bloom and Steen considered the US, which in comparison with other western countries, is one of the least forward looking in terms of child care policies. However, the 1990s has seen some improvement in terms of increased federal monies for various aspects of child care (for instance through tax credits). The US commitment to a "free market" economy means that discussions of child care, such as Bloom and Steen's, often emphasize the "child care industry" and the place of resource and referral centers, and commercial and workplace-based child care. In comparison, Canada has a more generous maternity leave policy and has tighter control over the regulation of child care (even for-profit/commercial care is subject to government regulation in Canada, whereas there is limited regulation in the US). Thus Truelove (Chapter 3) emphasized the regulation of child care, whereas Bloom and Steen did not. Canada has had a longer history of federal/provincial cost-sharing in terms of social services, and that there is a clear division of responsibilities among different levels of government regarding social services (but see Chapter 1, note 2). Thus, Truelove explored the ways that child care is manifest in a particular local setting: Metropolitan Toronto. Finally, Fincher provided a cross-national comparison of the literature on state interventions into the provision of child care. International comparisons allowed Fincher to highlight similarities and differences, and to assess the influence of contextual factors as well as the outcomes associated with different countries policies and provisions.

Cromley and Skelton dealt with child care at the intra-state/province level. A recurring theme in their chapters (and others) was that of spatial equity, especially in terms of rural–urban differences (also see Curtis, 1989; Phillips, 1991). Certainly, in their explorations of intra-state/province differences in the public provision of child care, Cromley and Skelton found significant spatial variations and noted that rural areas were more likely to be underserved, often chronically so. In her analysis of the provision of public full-day kindergarten, Cromley remarked that the towns in Connecticut's urban corridor, ranging from the southwestern part of the state (including Greenwich) to Hartford "played important roles in the development and adoption of early childhood programs in Connecticut and the US" (p. 54). In general, she noted that the availability of full-day and extended day kindergarten was greatest in urban communities. On the other hand, while some rural communities did offer such programs, others "remain a core area of resistance to the adoption of full-day kindergarten programs" (p. 60).

Skelton also found distinct rural–urban differences in provision in Ontario. He noted that the number of child care spaces was greatest in the urbanized area centering on Toronto, but in some areas availability was quite limited, especially in Northern Ontario. He pointed out that Ontario's Ministry of Community and Social Services acknowledges spatial variations in child care services and that one of its policy initiatives – Child Care Reform – is intended to redress spatial inequities in access to child care (that may, in part, have been generated by child care subsidies) by providing training, administration and financial support to under-supplied areas (but see Chapter 1, note 2).

Despite the general consensus that rural areas are often under-served regarding child care, there are few studies of child care in rural settings. Myers-Jones and Brooker-Gross provided an important exception with their case study of Blacksburg, Virginia, where the university, as one of the main employers, brings "a large well-educated middle-class population to a piece of Appalachia" (p. 82). Reflecting Cromley and Skelton's findings for rural areas, there are few child care options in Blacksburg. Moreover, Myers-Jones and Brooker-Gross thought Blacksburg might be "small enough to intuitively be a frictionless zone" (p. 86), so given this and constrained child care, they expected to find limited geographic sensitivity. However, one third of their respondents checked distances from home as an important factor in their choice of center. As Myers-Jones and Brooker-Gross point out, they may have picked up on "an artefact of small city life: low expectations of daily travel distances. . . . A larger urban center would have more child care options, more dispersed residential and employment patterns, and potentially a greater acceptance of the necessity for longer daily trips." (pp. 91–2).

Truelove (Chapter 7) provides an example of a larger urban center and alerted us to the importance of intra-urban differences in her study of the evolution of child care provision and journeys to child care in the context of municipal government policies in Metropolitan Toronto. She traced the spatial development of the three types of formal child care centers (non-profit, commercial and municipal) since 1971. All three types have grown, with non-profit displaying the highest rates, possibly encouraged by government policies favoring non-profit over commercial centers. However, she found clear inner city–suburban differences in these patterns. Generally, Truelove found that the inner city contained many non-profit and few commercial centers, whereas the opposite held for the suburbs. On the other hand, the growth rate of municipal centers was especially pronounced in suburban areas reflecting government responses to the mismatch between low provision and high demand. In Chapter 7, Truelove showed that subsidized children in Metropolitan Toronto had slightly shorter commutes than full-fee children. She also found that families paying full fees were traveling greater distances in the suburbs than in the inner city, whereas there was little spatial difference for subsidized children. Truelove suggests that this is because of the implicit government policy to encourage families with subsidies to attend their nearest center.

Fincher, Myers-Jones and Brooker-Gross, and Truelove indicated that geography is important even at a highly localized scale. Dyck and England developed this theme in their explorations of the everyday geographies of mothers living in suburbs. Dyck emphasized the centrality of neighborhood-based social networks in suburban Vancouver as one of the women's management strategies. She further argued that the suburbs contextualized how women define and reconstruct their identities and their neighborhood to facilitate the integration of their mothering and paid worker roles. She showed that the local context shaped the meaning and diversity of those practices, indicating that mothering and motherhood are not only socially constructed, but spatially situated. Similarly, England argued that the everyday experiences of mothers in suburban Columbus were often enmeshed in a complex web of localized social relations and networks. She suggested that mothers "create various coping strategies and attempt to alter their sociospatial systems so as to better negotiate their multiple roles" (p. 122).

CONCLUSION

The contributions to this book are indicative of the pluralist character of contemporary feminist geographies. For instance, the contributors tackled the issue of child care from a number of perspectives, employing a variety of approaches, data and research methods. A number engaged in policy analysis, others employed quantitative methods, while others used qualitative methods. At the same time, feminist geographies are apparent at a variety of scales and this is reflected in the contributions in that they dealt with child care at the national, regional, rural, metropolitan and local scales. And, of course, it is important to remember that these scales intersect with one another in that government (federal, provincial/state and municipal) policies filter down to the local level, that local activism can influence government policy, and local outcomes are embedded in wider spatial scales.

Child care continues to be an important issue in North America, often the topic of vigorous public debate. The contributors reflect the variety of child care arrangements in North America. Many of the authors dealt with formal child care, looking at the spatial variations in formal child care provision, and various aspects of trips to formal child care centers. However, the vast majority of children are cared for informally, and some of the authors looked at informal care as part of mothers' locally embedded management strategies. However, it is difficult to obtain information about the informal sector and, as Fincher and Truelove (Chapter 7) remark, future research should be aimed at a deeper understanding of informal care and its relation to formal care (see Mackenzie and Truelove, 1993, for an example of this).

Two themes were pervasive throughout the book. First is the place of child care in the gender division of labor. Whether at the scale of government policies or household strategies, child care is still constructed as women's work. Clearly the construction of femininities and male privilege continue as important structuring elements at different spatial scales. Second, is the myriad ways that a child's experience of child care is shaped by issues of class, "race"/ethnicity, residential location and so on. The current situation works to maintain distinctions that further marginalize particular groups of women, families and children. Not only are these issues for further research, but they are issues that need to be challenged along an array of spatial scales.

APPENDICES

APPENDIX I

One discriminant function was derived:

$$H = d_1Z_1 + d_2Z_2 + d_3Z_3 + d_4Z_4 + d_5Z_5$$

where:

H = score on the discriminant function
d_j = the weighting coefficients for the respective variables; and
Z_j = the standardized values of the five discriminating variables.

The analysis produced a set of standardized canonical coefficients:

Urban community type	0.79378
Median family income	0.47715
Labor force participation rate of mothers of young children	−0.17827
Presence of public school preschool program	0.16114
School district type, elementary through high school	0.15338

Each variable was significant at the 0.10 level.
The Chi-Square was 30.669 (5 d.f., p <00).

APPENDIX II

Regression results

Indicator	B	Beta	t	Significance
% white-collar employment	0.12	0.85	7.43	<.001
% labor force participation	–8.62	–0.70	–4.79	<.001
% labor force participation, women with children under age, 6	2.92	0.25	2.11	<.05
East (binary)	–76.31		–5.42	<.001
North (binary)	–59.49		–3.21	<.003
Constant	208.33		1.64	<.15
R	0.80			
R-squared	0.64			
Adjusted *R*-squared	0.60			
F	15.49			
Significance	<.001			

NOTES

CHAPTER 1 WHO WILL MIND THE BABY?

1 In both countries, as Bloom and Steen, Skelton, and Truelove (Chapters 2, 3 and 5) note, the first widespread government involvement in child care came during the 1930s and 1940s. In both cases, there had been limited involvement earlier, but this was almost entirely within a welfare context aimed at poor families (especially lone mothers). Beginning in 1933, the US federal government established child care centers and nursery schools under the New Deal legislation and the Works Progress Administration (WPA). Mobilization for the Second World War saw many of these programs disbanded, but in 1942, the Community Facilities Act (the Lanham Act) allowed funds to be channeled into regions where limited child care facilities impeded the war effort. The so-called Lanham Act Centers received 50 per cent of their funding from the federal government, with the states, municipalities and parents making up the rest. A similar level of cost-sharing was provided in Canada under the Dominion–Provincial War-Time Agreement of 1942. Federal government funding provided 50 per cent of the province's costs for child care programs allowing mothers to work in essential war industries. As Truelove (Chapter 3) notes, in practice, only Ontario and Québec were involved, partly because many other provinces were dominated by agriculture, which was not classified as an essential war industry. Both federal governments stressed the temporary nature of this funding, and at the end of the war, the funding was withdrawn. Both countries saw opposition and activism against this withdrawal (for example, the Day Nurseries and Day Care Parents Association in Toronto) that resulted in some centers briefly remaining open in Toronto, New York City, Washington, D.C., and California. Indeed, Zigler and Lang (1991) lament "[h]ad the system of child care established during the war remained in place, the [US] would now have a formal child care institution advantaged by decades of cultivation and experience" (1991: 35–6). However, Skelton (Chapter 5: 65) argues that Canada's "wartime program helped to shift public opinion from a vision of child care as a charity to that of a necessary service."

2 Writing is indeed a product of the time it is written! The period between the book going out for review and the point at which I am writing this footnote has been marked by tremendous change in social policy in Canada in general, and certain provinces (including Ontario) in particular. In recent years the CAP has come under tremendous pressure. The federal government recently announced plans to dissolve CAP and replace it (and the block grants for health and post-secondary education provided through Established Programs Financing) by the Canada

180

Health and Social Transfer beginning in 1996. The federal government had already capped its contributions to the three wealthiest provinces – Alberta, British Columbia and Ontario – in 1990 (whereas previously the amount had fluctuated, on a cost-sharing basis, according to the provincial contributions). Similar proposals for "financial constraint" have been made in the US, where AFDC (which also works on a federal/state cost-sharing basis) might be dismantled and replaced by block grants to the states. These block grants will be fixed amounts (that are smaller than the amounts of the previous programs) and may well decrease over time. The introduction of these block grants will reinforce rather than ameliorate provincial (or state) differences in wealth. These cuts will affect profoundly the quality of human services. This move enables the provinces (or states) to design their own welfare systems and experiment with welfare reform. Some commentators are concerned that funding for welfare and social assistance will receive less priority than health and education; and that where the CAP worked on a fairly loose definition of the basis of need, much tighter qualifications might now be introduced. Perhaps more worrisome is that the contours of provincial health and social programs will depend on not only provinces' or states' fiscal ability, but also their political orientation.

For quality licensed child care provision, the move to Canada Health and Social Transfers is potentially devastating. Indeed, Martha Friendly and Mab Oloman argue that in Canada "in the 1990s, child care has been characterized by cutbacks, changes in policy reflecting provincial governments' adoption of fiscally-driven social agendas, and shifts away from the concept of child care as an essential public service" (1995: 7). In Ontario, for example, the 1995 elections resulted in a Progressive Conservative government that ran on a platform of introducing deep cuts in the funding of social programs as a means to balancing the budget. As I write numerous social programs are being threatened: from social assistance to education. (Alberta has recently introduced similar cuts.)

In terms of child care, the proposed provincial cuts include slashing various pay subsidies to child care workers (including reversing legislation introduced under the previous provincial government that gave substantially larger subsidies to workers in non-profit centers) and suggestions that mothers on welfare provide home-based child care for other children. All of this means that there will be even greater spatial differences across Canada than before (for example, for the moment Québec's child care system is intact, as is British Columbia's), and, of course, it is a spatial diversity that is *not* driven by regional and local variation in children's and parents' child care needs!

In the very week that I am to send the final manuscript to Routledge, an internal report from the Ontario Ministry of Community and Social Services was leaked to the newspapers indicating that the provincial government is planning to phase out subsidies and introduce vouchers. The "market driven" argument is that vouchers increase parents' choice. However, the value of the proposed vouchers is well below the cost of formal care in some urban areas, and many low-income parents will find it very difficult to afford center care. The result will be that those centers dependent on subsidies (all municipal centers, many non-profit centers, and a smaller, but still substantial proportion of commercial centers) might be forced to close. Of course, hardest hit will be those centers in poorer, urban neighborhoods, but even those that cater to a range of children will be affected. If vouchers are introduced, child care advocates are concerned that formal care will become the domain of wealthy families, while poorer families will have to rely on the informal sector. Increased reliance on the informal sector further reinforces the gendered nature of child care as most informal care is provided by female relatives, neighbors and sitters. And, of course, with the increased paid employment of women there are few women available to provide informal care. (Most of the information in this footnote was provided by the Childcare Resource and Research Unit at the University of Toronto. Thanks to Martha Friendly and Chris Gehman for their help.)

3 According to Hägerstrand (1970) every person follows a daily (or monthly, or yearly) time-space path, along which there are stations (that are fixed in time and space) where the person stays for a period of time (for instance, the hours spent at work each day). Moving from one station to another is often hampered by a number of constraints. Hägerstrand identifies three types of constraint: "capability constraints" (for example: access to a car enables a person to travel a greater distance in a set time than people who are reliant on public transportation); "coupling constraints" (for example: the operating hours of services) and "authority constraints" (for example: limited spaces in a particular child care center).

CHAPTER 2 MINDING THE BABY IN THE UNITED STATES

1 This amount has stayed the same since 1971, and is well below the current average cost of full-time child care in the US of approximately $3,500 per year (Reeves, 1992).
2 The turnover rate is defined here as the percentage of individuals employed in an occupation who are not employed in that same occupation a year later.
3 These figures do not include the growing number of firms that provide child care assistance indirectly by allowing employees to pay for child care in pre-tax dollars under the flexible benefit provisions of the federal tax code.
4 However, this number is still in stark contrast to the Second World War, when more than 2,500 on-site child care centers that were established to help women enter the labor force.

CHAPTER 3 MINDING THE BABY IN CANADA

My thanks to Martha Friendly, Coordinator, Childcare Resource and Research Unit, Centre for Urban and Community Studies, University of Toronto, for her comments on an earlier version of this chapter.

1 Foster care and other special forms of care are not included here; child care refers to care of children who live with their parents or guardians. Care of a child whose mother does not work outside the home is not usually called child care; in-home care by a parent usually means care by the father while the mother works (this is indicative of the attitude that parenting is mostly mothers' work).
2 The Mother's Allowance Act was passed province by province; Québec did not enact it until the 1930s.
3 In Ontario, only 1,147 children received subsidized child care in 1950 (Krashinsky, 1977). The Day Nurseries Act was not changed significantly for twenty years; during this time there was little growth in the number of publicly provided child care spaces, and funding was cut back for several centers.

CHAPTER 5 CHILD CARE SERVICES IN ONTARIO

1 This is the regionalization adopted by the Ministry of Community and Social Service. Details can be found in Ontario MCSS (1984: Table F7.1).
2 Where there are large correlations between groups of variables, they were combined using principle components analysis. However, it was found that the best results were associated with models employing individual, non-correlated variables rather than those with combination variables.

CHAPTER 6 THE JOURNEY TO CHILD CARE IN A RURAL AMERICAN SETTING

We thank several students for their assistance in the design and data collection stages of this project: Tony Benger, Jeffery Bunono, Dominic Mauriello, Gus Colom and Mike Martin.

CHAPTER 7 THE LOCATIONAL CONTEXT OF CHILD CARE CENTERS IN METROPOLITAN TORONTO

I wish to acknowledge the funding support of the Child Care Initiatives Fund, Health and Welfare Canada (4774–6–91/3(652)).

1 In December 1993, the Canadian labor force participation rate for mothers with at least one child under 6 years was 64.1 per cent (Statistics Canada Catalogue 71–001, December, 1993). The 1991 Census of Canada (Statistics Canada, 1992) shows that there were 155,345 children under 6 years old in Metropolitan Toronto. If one assumes that the Canadian labor force participation rate applied to Metropolitan Toronto, then approximately 100,000 children have mothers in the paid labor force. If half of those mothers in the paid labor force do not work full-time, or prefer informal child care, there are approximately 50,000 children whose parents prefer child care centers.
2 The size of centers varies widely but they are much smaller than centers in the US. Regardless of ownership type, they average a capacity of fifty-eight to sixty children. Each local municipality also has a range of center sizes, from a low of ten to a high of over 100 spaces. Comparisons with 1971 data indicate that the average capacity has not changed that much over time.
3 The travel to child care centers analysis draws on data originally collected in 1984 for a broader project (Truelove, 1989). At that time, Metropolitan Toronto had an implicit policy of selecting the centers that children with subsidies would attend. These centers were usually near their homes. Moreover, these data also show short distances even at a time when there were relatively fewer centers than there are today.
4 Compusearch Market and Social Research Limited (now Compusearch Micro-Marketing Data and Systems Limited) provided the enumeration areas and grid references for the EA centroids for approximately 1,300 of the addresses in the sample. Information for the remaining 300 addresses was obtained by hand from grid reference maps.
5 Infant care is most expensive, so those centers that cater only to infants have higher prices. However, this price difference affects only a few of the centers in the population and only one in the address sample for this study. In that case, the

price of infant care was multiplied by a factor of 0.80, after observing the price differentials for several centers with two branches serving different age groups.

6 The Chi Square test is a test of statistical significance; the frequency distributions for two variables are compared.

7 However, a *t*-test (a test of statistical significance for the difference between two samples with ratio-level data) shows no significant difference between the two mean distances (3.1 km for full-fee and 2.8 km for subsidized children).

8 The travel patterns of full-fee versus subsidized children was also examined by type of center. The mean distance traveled to commercial centers is 3.2 km for full-fee and 3.4 km for subsidized children. For non-profit centers, full-fee distances average 3.1 km, but only 2.5 km for subsidized children. However, for all types of centers greater *proportions* of subsidized children than full-fee children travel very short distances.

9 The Etobicoke sub-sample of four centers includes a municipal center that is next door to a public housing complex. Thus, it is not surprising that 34.5 per cent of Etobicoke children attending a child care center travel 0.5 km or less.

CHAPTER 8 MOTHERS, WIVES, WORKERS

I wish to thank the women and personnel managers whom I interviewed, they are identified by pseudonyms. Thanks also to Audrey Glasbergen who patiently typed and retyped earlier versions of the manuscript.

1 One of the most "commonsense" explanations as to why women have shorter work-trips than men is related to women's domestic responsibilities and activities. Curiously, there is mixed evidence regarding the presence of children as an explanation for women's shorter commutes. Some studies find that the presence of children means that women work closer to home (Pickup, 1984; Preston *et al.*, 1993; Singell and Lillydahl, 1986); while others conclude that mothers have longer or similar commutes to those of women without children (Gordon *et al.*, 1989; Hanson and Johnston, 1985; Johnston-Anumonwo, 1992).

2 The practice of families depending on older children or other family members for child care, domestic labor or wages, was quite common up until the Second World War (Brodkin-Sacks, 1984; Roberts, 1986).

CHAPTER 9 MOTHER OR WORKER?

1 See for example, Wekerle (1980); special issue of *Urban Geography*, 9(2), 1988; Little *et al.* (1988).

2 A range of viewpoints on these issues can be found in Reinharz (1992) and Harding (1987).

3 I conducted this participant observation as a mother of two young children, so was an "insider" in the research situation. Some of the advantages and disadvantages of an insider perspective are discussed in Aguilar (1981), although feminist scholars have tended to emphasize the value of having experiences related to a research topic, provided the researcher is reflexive about her positioning within the research process. In this research, advantages included unproblematic access to a variety of settings, knowledge of the language and "commonsense" understandings of many mothers with young children in the study area which facilitated communication, and opportunity to discuss at length, and informally, a variety of child-related issues.

4 The names used in this chapter to identify particular women and children are pseudonyms.
5 Shared interests and everyday experiences, such as between mothers in caring for their children, have been shown to be formative of women's values and ideas, and a basis for common viewpoints (Dyck, 1990; Gullestad, 1984; Ruddick, 1989).

CHAPTER 10 THE STATE AND CHILD CARE

The research from which this chapter is drawn is funded by the Australian Research Council. I am grateful for its support, and also for the excellent library-foraging efforts of Jillian Oldfield.

REFERENCES

Action for Children (1989) *Child Care in Columbus: An Economic and Child Care Policy Paper*, Columbus: Action for Children.

Adams, C. T. and Winston, K. T. (1980) *Mothers at Work: Public Policies in the United States, Sweden and China*, New York and London: Longman.

Aguilar, J. L. (1981) "Insider research: An ethnography of a debate," in D. A. Messerschmidt (ed.) *Anthropologists at Home in North America: Methods and Issues in the Study of One's Own Society*, Cambridge: Cambridge University Press.

Andre, T. and Neave, C. (1992) *The Complete Canadian Day Care Guide*, Toronto: McGraw-Hill Ryerson.

Anyon, J. (1983) "Intersections of gender and class: Accommodation and resistance by working-class and affluent females to contradictory sex-role ideologies," in S. Walker and L. Barton (eds.) *Gender, Class and Education*, Lewes, Sussex: The Falcon Press.

Armitage, A. (1988) *Social Welfare in Canada*. Toronto: McClelland and Stewart.

Badelt, C. (1991) "Austria: Family work, paid employment, and family policy," in S. Kamerman and A. Kahn (eds.) *Child Care: Parental Leave and the Under 3s*, New York: Auburn House.

Bakan, A. B. and Stasiulis, D. K. (1995) "Making the match: Domestic placement agencies and the racialization of women's household work," *Signs* 20(2): 303–35.

Baker, M. (1987) *Child Care Services in Canada*, Background paper for parliamentarians No. 122, Ottawa: Supply and Services.

Barnes, T. (1987) "Homo economicus, physical metaphors, and universal models in economic geography," *The Canadian Geographer* 31: 299–308.

Barnhorst, R. and Johnson, L. C. (eds.) (1991) *The State of the Child in Ontario*, Toronto: Oxford University Press.

Barnsley, J. (1988) "Feminist action, institutional reaction," *Resources for Feminist Research* 17(3): 18–21.

Beach, J. (1992) "A comprehensive style of child care," cited in M. Friendly (1994) *Child care Policy in Canada: Putting the Pieces Together*, Don Mills, Ontario: Addison-Wesley Publishers.

Beach, J., Friendly, M. and Schmidt, L. (1993) *Work-related Child Care in Context: A Study of Work-related Child Care in Canada*, Occasional Paper No. 3, Toronto: Child Care Resource and Research Unit, University of Toronto, Centre for Urban and Community Studies.

Bean, F. D. and Tienda, M. (1987) *The Hispanic Population of the United States*, New York: Russell Sage Foundation.

Becker, G. (1981) *A Treatise on the Family*, Cambridge, Massachusetts: Harvard University Press.

Benito-Alonso, M. A. and Devaux, P. (1981) "Location and size of day nurseries: A multiple goal approach," *European Journal of Operations Research* 6: 195–8.

Bennett, L. (1991) "Women, exploitation and the Australian child-care industry: Breaking the vicious circle," *The Journal of Industrial Relations* 33(1): 20–40.

Bianchi, S. M. and Spain, D. (1986) *American Women in Transition*, New York: Russell Sage.

Blau, D. and Robins, P. (1988) "Child-care costs and family labor supply," *Review of Economics and Statistics* 70: 374–81.

Blau, F. D. and Ferber, M. A. (1992) *The Economics of Women, Men and Work*, Englewood Cliffs, New Jersey: Prentice-Hall.

Blewett, N. (1989) "Health, human services, the economy and social justice," in P. Saunders and A. Jamrozik (eds.) *Social Policy in Australia: What Future for the Welfare State?* Sydney: University of New South Wales, Social Welfare Research Centre Reports and Proceedings No. 79.

Bloom, D. E. (1986) "Fertility timing, labor supply disruptions, and the wage profiles of American women," *Proceedings of the American Statistical Association, Social Statistics Section*: 49–63.

Bloom, D. E. and Steen, T. P. (1990) "The labor force implications of expanding the child care industry," *Population Research and Policy Review* 19: 25–44.

Bloom, D. E. and Trahan, J. T. (1986) *Flexible Benefits and Employee Choice*, New York: Pergamon Press.

Bloom, D. E. and Trahan, J. T. (1987) "The labor force implications of expanding the child care industry," *Population Research and Policy Review* 19: 25–44.

Blumen, O. (1994) "Gender differences in the journey to work," *Urban Geography* 15(3): 223–45.

Boles, J. (1989) "A policy of our own: Local feminist networks and social services for women and children," *Policy Studies Review* 8(3): 638–47.

Bondi, L. (1992) "Gender and dichotomy," *Progress in Human Geography* 16: 98–104.

Bookman, A. and Morgen, S. (eds.) (1988) *Women and the Politics of Empowerment*, Philadelphia: Temple University Press.

Borchorst, A. (1990) "Political motherhood and child care policies: A comparative approach to Britain and Scandinavia," in C. Ungerson (ed.) *Gender and Caring*, New York: Harvester Wheatsheaf.

Boulton, M. G. (1983) *On Being a Mother*, London: Tavistock.

Bowlby, S. (1986) "The place of gender in locality studies," *Area* 18: 327–31

Bowlby, S. (1990) "Women, work and the family: Control and constraints," *Geography* 75: 17–26.

Brennan, D. (1989) "Private vs public: Child care revisited," *Current Affairs Bulletin* 65(12): 27–28.

Brennan, D. (1992) "Children's services: Debates and dilemmas," *Impact* 22(3): 10–11.

Brodkin-Sacks, K. (1984) "Generations of working-class families," in K. Brodkin-Sacks and D. Remy (eds.) *My Troubles Are Going to Have Trouble with Me: Everyday Trials and Triumphs of Women Workers*, New Brunswick: Rutgers University Press.

Bronfenbrenner, U. (1992) "Child care in the Anglo-Saxon mode," in M. E. Lamb, K. J. Sternberg, C. P. Hwang and A. G. Broberg (eds.) *Child Care in Context: Cross-cultural Perspectives*, Hillsdale, New Jersey: Lawrence Erlbaum Associates.

Brooker-Gross, S. R. and Maraffa, T.A (1985) "Commuting distance among nonmetropolitan university employees," *Professional Geographer* 39: 309–17.

Brown, L. A. (1981) *Innovation Diffusion: A New Perspective*, London and New York: Methuen.

REFERENCES

Brown, L. A., Williams, F.B., Youngman, C.E., Holmes, J. and Walby, K. (1972) *Day Care Centers in Columbus: A Locational Strategy*, Discussion Paper No. 26, Department of Geography, Ohio State University.

Burke, M. A., Crompton, S., Jones, A. and Nessner, K. (1994) "Caring for children," in C. McKie and K. Thompson (eds.) *Canadian Social Trends: A Canadian Studies Reader (Volume 2)*, Toronto: Thompson Educational Inc.

Canada (1986) *Report of the Task Force on Child Care*, Ottawa: Status of Women.

Canada Dominion Bureau of Statistics (1953) *Ninth Census of Canada, 1951 (Vol. 1): Population by Specific Age Groups and Sex for Counties and Census Divisions*. Ottawa: Queen's Printer.

Canada Dominion Bureau of Statistics (1962) *1961 Census of Canada: Population*, Catalogue 92–713. Ottawa: Queen's Printer.

Canadian Advisory Council on the Status of Women (1984) *Day Care in Canada: A Background Paper*, Ottawa.

Carroll, B.W. (1989) "Administrative devolution and accountability: The case of the non-profit housing program," *Canadian Public Administration/Administration publique du Canada* 32: 345–66.

Carter, P. (1987) *A Report on Child Care Services in Ontario*, Social Assistance Review Committee Report Document No. 37, Toronto: Ontario MCSS.

Casper, L. (1995a) "Child care costs greater burden for the poor," (posted on the Internet: www.census.gov).

Casper, L. (1995b) "What does it cost to mind our preschoolers?" (posted on the Internet: www.census.gov).

Chamber of Commerce (1986) *Economic Profile: An Overview of Life, Business and Government in the Region*, Coquitlam, Port Coquitlam, Port Moody: Chamber of Commerce.

Chouinard, V. and Fincher, R. (1987) "State formation in capitalist society," *Antipode* 19: 329–53.

Clark, W. A. V. and Hosking, P. L. (1986) *Statistical Methods for Geographers*, New York: John Wiley and Sons.

Clerkx, L. and Van Ijzendoorn, M. (1992) "Child care in a Dutch context: On the history, current status and evaluation of nonmaternal child care in the Netherlands," in M. Lamb, K. Sternberg, C. Hwang and A. Broberg (eds.) *Child Care in Context*, Hillsdale, New Jersey: Lawrence Erlbaum Associates.

Cohen, M. G. (1993) "Social policy and social services," in R. R. Pierson, M. G. Cohen, P. Bourne and P. Masters *Canadian Women's Issues: Strong Voices*, Toronto: Lorimer and Company.

Committee on the Infant and Preschool Child (1931) *White House Conference on Child Health and Protection: Nursery Education*, New York: The Century Co.

Community Information Centre of Metropolitan Toronto (various years) *Day Care*. Toronto: Annual Public Document.

Community Services Department (1990) *Face of the Future: Comprehensive Review of Childcare Report*, Metropolitan Toronto and Toronto Area Office: Ministry of Community and Social Services.

Community Services Department (1992) *Service Plan for Child Care Services 1992–93*, Metropolitan Toronto and Toronto Area Office: Ministry of Community and Social Services.

Connecticut Early Childhood Education Council (1983) *Report on Full-Day Kindergarten*, Hartford, Connecticut: Connecticut Early Childhood Education Council.

Corsaro, W. and Emiliani, F. (1992) "Child care, early education and children's peer culture in Italy," in M. Lamb, K. Sternberg, C. Hwang and A. Broberg (eds.) *Child Care in Context*, Hillsdale, New Jersey: Lawrence Erlbaum Associates.

REFERENCES

Cromley, E. (1987) "Locational problems and preferences in preschool child care," *Professional Geographer* 39: 309–17.

Cryan, J. R., Sheehan, R., Wiechel, J. and Bandy-Hedden, I. G. (1992) "Success outcomes of full-day kindergarten: More positive behavior and increased achievement in the years after," *Early Childhood Research Quarterly* 7: 187–93.

Curtis, S. (1989) *The Geography of Public Welfare Provision*, London and New York: Routledge.

d'Abbs, P. (1991) *Who Helps? Support Networks and Social Policy in Australia*, Melbourne: Australian Institute of Family Studies, Monograph Number 12.

Dally, A. (1982) *Inventing Motherhood*, London: Burnett Books.

Darian, J. C. (1975) "Convenience of work and the job constraint of children," *Demography* 12: 245–55.

David, M. (1984) "Women, family and education," in S. Acker, J. Megarry, S. Nisbet and E. Hoyle (eds.) *World Yearbook of Education 1984: Women and Education*, London: Kogan Page.

Deskins, D. (1973) "Residence–workplace interaction vectors for the Detroit Metropolitan Area: 1953–1965," in M. Albaum (ed.) *Geography and Contemporary Issues*, New York: Wiley.

Dex, S. and Shaw, L. (1986) *British and American Women at Work*, London: Macmillan.

Doyal, L. and M. A. Elston. (1986) "Women, health and medicine," in V. Beechey and E. Whitelegg (eds.) *Women in Britain Today*, Milton Keynes: Open University Press.

Droogleever Fortuijn, J. and Karsten, L. (1989) "Daily activity patterns of working parents in the Netherlands," *Area* 21: 365–76.

Dyck, I. (1989) "Integrating home and wage workplace: Women's daily lives in a Canadian suburb," *The Canadian Geographer* 33(4): 329–41.

Dyck, I. (1990) "Space, time, and renegotiating motherhood: An exploration of the domestic workplace," *Environment and Planning D: Society and Space* 8: 459–83.

England, K. (1993a) "Suburban pink-collar ghettos: The spatial entrapment of women?" *Annals of the Association of American Geographers* 83: 225–42.

England, K. (1993b) "Changing suburbs, changing women: Geographic perspectives on suburban women and suburbanization," *Frontiers: A Journal of Women Studies* 14(1): 24–43.

England, K. and Steill, B. (forthcoming) "'They think you're as stupid as your English is': Constructing the national identities of foreign domestic workers in Toronto," *Environment and Planning A*.

Ericksen, J. A. (1977) "An analysis of the journey to work for women," *Social Problems* 24: 428–35.

Fagnani, J. (1983) "Women's commuting patterns in the Paris region," *Tijdschrift Voor Economische en Sociale Geografie* 74: 12–24.

Farley, R. and Allen, W. R. (1987) *The Colorline and the Quality of Life in America*, New York: Russell Sage Foundation.

Ferguson, E. (1991) "The child-care crisis: Realities of women's caring," in C. Baines, P. Evans and S. Neysmith (eds.) *Women's Caring: Feminist Perspectives on Social Welfare*, Toronto: McClelland and Stewart Inc.

Fernandez, J. (1990) *The Politics and Reality of Family Care in Corporate America*, Lexington, Massachusetts: Lexington Books.

Fincher, R. (1989) "Class and gender relations in the local labor market and the local state," in J. Wolch and M. Dear (eds.) *The Power of Geography*, Boston: Unwin Hyman.

Fincher, R. (1991) "Caring for workers' dependents: Gender, class and local state practice in Melbourne," *Political Geography Quarterly*, 10: 356–81.

Fincher, R. (1993) "Women, the state and the life course in urban Australia," in C. Katz and J. Monk (eds.) *Full Circles: Geographies of Women over the Life Course*, London and New York: Routledge.

Fine, B. (1992) *Women's Employment and the Capitalist Family*, London and New York: Routledge.

Fodor, R. (1978) "Day-care policy in France and its consequences for women: A study of the Metropolitan Paris area," *International Journal of Urban and Regional Research* 2: 463–81.

Fooks, C. (1987) *Child Care in Ontario: The Debate Over Commercial Day Care*, Legislative Research Services, Current Issues Paper No. 70, Toronto: The Queen's Printer for Ontario.

Franzway, S., Court, D. and Connell, R. (1989) *Staking a Claim*, Sydney: Allen and Unwin.

Friendly, M. (1992) "Moving towards quality child care: Reflections on child care policy in Canada," *Canadian Journal of Research in Early Childhood Education* 3: 123–32.

Friendly, M. (1994) *Child Care Policy in Canada: Putting the Pieces Together*, Don Mills, Ontario: Addison-Wesley Publishers.

Friendly, M. and O'Neill, P. (1983) *Toward a coordinated child care system in North York*, Children's Services Committee of the North York Inter-Agency Council.

Friendly, M. and Oloman, M. (1995) "Child care at the centre: Child care and the social, economic and political agenda in the 1990s," paper presented at the Social Welfare Policy Conference, Vancouver, BC.

Friendly, M., Mathien, J. and Willis, T. (1987) *Child Care – What the Public Said: An analysis of the transcripts of the public hearings held across Canada by the Parliamentary Special Committee on Child Care*, Ottawa: Canadian Day Care Advocacy Association.

Friendly, M., Rothman, L. and Oloman, M. (1991) *Child Care for Canadian Children and Families*, Occasional Paper No. 31, Childcare Resource and Research Unit, Centre for Urban and Community Studies, University of Toronto.

Fromberg, D. P. (1987) *The Full-Day Kindergarten*, New York: Teachers College Press, Columbia University.

Fullerton, H. N. (1993) "Another look at the labor force," *Monthly Labor Review* 116: 31–40.

Gardner, R. (1984) *Day Care in Ontario*, Legislative Research Services, Current Issues Paper No. 23, Toronto: The Queen's Printer for Ontario.

Gardner, R. K. (1986) *Kindergarten Programs and Practices in Public Schools*, Arlington: Educational Research Service Inc.

Gatfield, R. and Griffin, V. (1990) *Shift Workers and Child Care: A Study of the Needs of Queensland Nurses*, Canberra: Australian Government Publishing Service.

Gifford, J. (1992) "The social wage and child care," *Impact* 22(10): 15–17.

Glenn, E. N. (1992) "From servitude to service work: Historical continuities in the racial division of paid reproductive labor," *Signs* 19(1): 1–43.

Goelman, H. (1992) "Day care in Canada," in M. E. Lamb, K. Sternberg, C. Hwang and A. Broberg (eds.) *Child Care in Context*, Hillsdale, New Jersey: Lawrence Erlbaum Associates.

Gordon, P., Kumar, A. and Richardson, H. (1989) "Gender differences in metropolitan travel behavior," *Regional Studies* 23: 499–510.

Gormley, W. T. (1995) *Everybody's Children: Child Care as a Public Problem*, Washington, D.C.: The Brookings Institution.

Gregory, D. (1981) "Human agency and human geography," *Transactions, Institute of British Geographers* N.S. 6: 1–18.

Gregson, N. and Lowe, M. (1994) *Servicing the Middle Classes: Class, Gender and Waged Domestic Labour in Contemporary Britain* London and New York: Routledge.

Guberman, N. (1990) "The family, women and caregiving," in V. Dhruvarajan (ed.) *Women and Well-Being/Les femmes et le mieux-être*, Montréal and Kingston: McGill-Queen's University Press.

Gullestad, M. (1984) *Kitchen-Table Society: A Case Study of the Family Life and Friendships of Young Working-Class Mothers in Urban Norway*, Oslo: Universitetsforlaget.

Haas, L. (1992) *Equal Parenthood and Social Policy*, Albany: State University of New York Press.

Hägerstrand, T. (1970) "What about people in regional science?" *Papers and Proceedings of the Regional Science Association* 27: 7–21.

Hanson, S. and Johnston, I. (1985) "Gender differences in work-trip length: Explanations and implications," *Urban Geography* 6: 193–219.

Hanson, S. and Pratt, G. (1988) "Reconceptualising the links between home and work in urban geography," *Economic Geography* 64: 299–321.

Hanson, S. and Pratt, G. (1990) "Geographic perspectives on the occupational segregation of women," *National Geographic Research* 6(4): 376–99.

Hanson, S. and Pratt, G. (1991) "Job search and the occupational segregation of women," *Annals of the Association of American Geographers* 81: 229–53.

Hanson, S. and Pratt, G. (1992) "Dynamic dependencies: A geographic investigation of local labor markets," *Economic Geography* 68: 373–405.

Harding, S. (ed.)(1987) *Feminism and Methodology*, Bloomington and Indianapolis: Indiana University Press and Milton Keynes: Open University Press.

Harvey, D. (1973) *Social Justice and the City*, Oxford: Blackwell.

Harvey, D. (1992) "Social justice, postmodernism and the city," *International Journal of Urban and Regional Research* 16: 588–601.

Hayes, C. D., Palmer, J. L. and Zaslow, M. J. (eds.) (1990) *Who Cares for America's Children: Child Care Policy for the 1990s*, Washington, D.C.: National Academy Press.

Health and Welfare Canada (1991) *Status of Day Care in Canada (Annual)*. Ottawa: National Day Care Information Centre, Health and Welfare Canada.

Hiebert, E. H. (1988) "Introduction," *The Elementary School Journal* 89: 115–17.

Hochschild, A. with A. Machung (1989) *The Second Shift*, New York: Avon.

Hodgson, M. J. and Doyle, P. (1978) "The location of public services considering the mode of travel," *Socio-economic Planning Science* 12: 49–54.

Hofferth, S. L. and Phillips, D. A. (1987) "Child care in the United States, 1970 to 1995," *Journal of Marriage and the Family* 49: 559–71.

Hofferth, S. L., Brayfield, A., Deich, S. and Holcomb, P. (1991) *National Child Care Survey, 1990*, Washington, D.C.: Urban Institute Press.

Hurl, L. (1984) "Privatized social service systems: Lessons from Ontario children's services," *Canadian Public Policy/Analyse de politiques* 10: 395–405.

Hurl, L. and Tucker, D. (1986) "Limitations of an act of faith: An analysis of the Macdonald Commission's stance on social services," *Canadian Public Policy/Analyse de politiques* 12: 606–21.

Hwang, C. and Broberg, A. (1992) "The historical and social context of child care in Sweden," in M. Lamb, K. Sternberg, C. Hwang and A. Broberg (eds.) *Child Care in Context*, Hillsdale, New Jersey: Lawrence Erlbaum Associates.

Ismael, J. (ed.) (1987) *The Canadian Welfare State: Evolution and Transition*, Edmonton: University of Alberta Press.

Jaggar, A. (1983) *Feminist Politics and Human Nature*, Totowa, New Jersey: Rowman and Allanheld.

Johnson, L. (1977) *Who Cares?* Toronto: Social Planning Council of Metropolitan Toronto.

Johnston-Anumonwo, I. (1992) "The influence of household type on gender differences in work trip distance," *The Professional Geographer* 44: 161–9.

Kahn, A. J. and Kamerman, S. B. (1987a) "Local child care initiatives," in A. J. Kahn and S. B. Kamerman, *Child Care: Facing the Hard Choices*, Dover, Massachusetts: Auburn House Publishing Company.

Kahn, A. J. and Kamerman, S. B. (1987b) *Child Care: Facing the Hard Choices*, Dover, Massachusetts: Auburn House Publishing Company.

Kamerman, S. (1989) "Child care, women, work, and the family: An international overview of child care services and related policies," in J. Lande, S. Scarr and N. Gunzerhauser (eds.) *Caring for Children: Challenge to America*, Hillsdale, New Jersey: Lawrence Erlbaum Associates.

Kamerman, S. (1991) "Child care policies and programs: An international overview," *Journal of Social Issues* 47(2): 179–96.

Kamerman, S. B. and Kahn, A. J. (1989) "Child care and privatization under Reagan," in S. B. Kamerman and A. J. Kahn (eds.) *Privatization and the Welfare State*, Princeton: Princeton University Press.

Kanaroglou, P. S. and Rhodes, S. A. (1990) "The demand and supply of child care: The case of the city of Waterloo, Ontario," *The Canadian Geographer* 34: 209–24.

Karweit, N. (1988) "Quality and quantity of learning time in preprimary programs," *The Elementary School Journal* 89: 119–33.

Kelly, A. H. and Harbison, W. A. (1970) *The American Constitution: Its Origins and Development*, New York: W. W. Norton and Company.

Kipnis, B. A. and Mansfeld, Y.(1986) "Work-place utilities and commuting patterns: Are they class or place differentiated?" *Professional Geographer* 38: 160–9.

Kissman, K. (1991) "Women caregivers, women wage earners: Social policy perspectives in Norway," *Women's Studies International Forum* 14(3): 193–99.

Kitchen, B. (1990) "Family policy," in M. Baker (ed.) *Families: Changing Trends in Canada*, 2nd edn, Toronto: McGraw-Hill Ryerson.

Klein, A. (1992) *The Debate over Child Care, 1969–1990*, Albany: SUNY Press.

Kramer, R. (1981) *Voluntary Agencies in the Welfare State*, Berkeley: University of California Press.

Krashinsky, M. (1977) *Day Care and Public Policy in Ontario*, Toronto: Ontario Economic Council.

Labour Canada (1986) *Women in the Labour Force*, 1985–1986 edn, Ottawa: Labour Canada.

Labour Canada (1990) *Labour Standards Legislation*, Ottawa: Ministry of Supply and Services.

Lamb, M. and Sternberg, K. (1992) "Sociocultural perspectives on nonparental child care," in M. Lamb, K. Sternberg, C. Hwang and A. Broberg (eds.) *Child Care in Context: Cross-Cultural Perspectives*, Hillsdale, New Jersey: Lawrence Erlbaum Associates.

Lamb, M., Sternberg, K., Hwang, C. and Broberg, A. (eds.) (1992) *Child Care in Context: Cross-Cultural Perspectives*, Hillsdale, New Jersey: Lawrence Erlbaum Associates.

Lang, V. (1974) *The Service State Emerges in Ontario: 1945–1973*, Toronto: Ontario Economic Council.

Lazar, I., Darlington, R., Murray, H., Royce, J. and Snipper, A. (1982) "Lasting effects of early education," *Monographs of the Society for Research in Child Development* 47: Serial No. 194.

Le Grand, J. and Robinson, R. (eds.) (1984) *Privatisation and the Welfare State*, London: Allen and Unwin.

Ledoux, G. (1987) *Child Care in Canada*, Current Issue Review 87–11E, Ottawa: Library of Parliament.

Léger, H. and Rebick, J. (1993) *The NAC Voter's Guide*, Québec: Voyageur.

Leira, A. (1990) "Coping with care: Mothers in a welfare state," in C. Ungerson (ed.) *Gender and Caring*, New York: Harvester Wheatsheaf.

Lero, D. S. (1981) *Factors Influencing Parents' Preferences for and Use of Alternative Child Care Arrangements for Pre-school Age Children*, Final Report of a Project funded by Health and Welfare Canada.

Lero, D. S. (1985) *Parents' Needs, Preferences and Concerns about Child Care: Case Studies of 336 Canadian Families*, Final Report of a Study commissioned by the Task Force on Child Care, Ottawa: Task Force on Child Care.

Lero, D. S. and Kyle, I. (1991) "Families and children in Ontario," in L. C. Johnston and D. Barnhorst (eds.) *Children, Families and Public Policy in the 1990s*, Toronto: Thompson Educational Publishing Inc.

Lester, C. W. (1990) *Geographical Distribution of 1988, 1989, and 1990 Extended Day Kindergarten Grant Program and Non-Grant Programs*, Hartford, Connecticut: State Department of Education.

Lewis, J. and Astrom, G. (1992) "Equality, difference and state welfare: Labor market and family policies in Sweden," *Feminist Studies* 18(1): 59–87.

Ley, D.(1980) "Liberal ideology and the postindustrial city,' *Annals Association of American Geographers* 70: 238–58.

Lilley, I. M. (1967) *Friedrich Froebel: A Selection from His Writings*, Cambridge: Cambridge University Press.

Little, J., Peake, L. and Richardson, P. (eds.) (1988) *Women in Cities: Gender and the Urban Environment*, New York: New York University Press.

Lowe, M. and Gregson, N. (1989) "Nannies, cooks, cleaners, au pairs . . . new issues for feminist geography?" *Area* 21: 415–17.

Mackenzie, S. (1987) "Neglected spaces in peripheral places: Homeworkers and the creation of a new economic centre," *Cahiers de Géographie du Québec* 31(83): 247–60.

Mackenzie, S. (1989) "Restructuring the relations of work and life: Women as environmental actors, feminism as geographic analysis," in A. Kobayashi and S. Mackenzie (eds.) *Remaking Human Geography*, Boston: Unwin Hyman.

Mackenzie, S. and Rose, D. (1983) "Industrial change, the domestic economy and home life," in J. Anderson, S. Duncan and R. Hudson (eds.) *Redundant Spaces in Cities and Regions*, New York: Academic Press.

Mackenzie, S. and Truelove, M. (1993) "Changing access to public and private services: Non-family childcare," in L. S. Bourne and D. F. Ley (eds.) *The Changing Social Geography of Canadian Cities*, Montréal and Kingston: McGill-Queen's University Press.

Magnusson, W., Carroll, W. K., Doyle, C., Langer, M. and Walker, R. B. J. (eds.) (1984) *The New Reality: The Politics of Restraint in British Columbia*, Vancouver: New Star Books.

Maraffa, T. A. and Brooker-Gross, S. R. (1984) "Aspects of the journey to work within a small city laborshed," *Urban Geography* 5: 178–86.

Marchak, P. (1984) "The new economic reality: Substance and rhetoric," in W. Magnusson, W. K. Carroll, C. Doyle, M. Langer and R. Walker (eds.) *The New Reality: The Politics of Restraint in British Columbia*, Vancouver: New Star Books.

Mark-Lawson, J., Savage, M. and Warde, A. (1985) "Gender and local politics: Struggles over welfare policies," in L. Murgatroyd, M. Savage, D. Shapiro, J. Urry, S. Walby, A. Warde with J. Mark-Lawson *Localities, Class and Gender*, London: Pion.

Martensson, S. (1977) "Childhood interaction and temporal organization," *Economic Geography* 53: 99–125.

Martinez, S. (1989) "Child care and federal policy," in J. Lande, S. Scarr and N. Gunzenhauser (eds.) *Caring for Children: Challenge to America*, Hillsdale, New Jersey: Lawrence Erlbaum Associates.

McDaniel, S.M. (1993) "The changing Canadian family," in S. Burt, L. Code, and L. Dorney (eds.) *Changing Patterns: Women in Canada*, Toronto: McClelland and Stewart.

McDowell, L. (1993) "Space, place and gender relations, Part 1: Feminist empiricism and the geography of social relations," *Progress in Human Geography* 17: 157–79.

McIntyre, E. (1979) *The Provision of Day Care in Ontario: Responsiveness of Provincial Policy to Children at Risk because their Mothers Work*, D.S.W. Thesis, University of Toronto.

McLafferty, S. and Preston, V. (1991) "Gender, race, and commuting among service sector workers," *Professional Geographer* 43: 1–15.

McLafferty, S. and Preston, V. (1992) "Spatial mismatch and labor market segmentation for African-American and Latina women," *Economic Geography* 68: 406–31.

McLanahan, S. and Garfinkel, I. (1993) "Single motherhood in the United States: Growth, problems, and policies," in J. Hudson and B. Galaway (eds.) *Single-Parent Families: Perspectives on Research and Policy*, Toronto: Thompson Educational Publishing Inc.

Melhuish, E. C. and Moss, P. (eds.) (1991a) *Day Care for Young Children: International Perspectives*, London and New York: Routledge.

Melhuish, E. C. and Moss, P. (1991b) "Current and future issues in policy and research," in E. C. Melhuish and P. Moss (eds.) *Day Care for Young Children: International Perspectives*, London and New York: Routledge.

Meyer, J. W. (1975) *Diffusion of an American Montessori Education*, Research Paper No. 160, Chicago: Department of Geography, The University of Chicago.

Michelson, W. (1983) "The logistics of maternal employment: Implications for women and their families," *Child in the City Report No. 18*, Centre for Urban and Community Studies, University of Toronto.

Michelson, W. (1985) *From Sun to Sun*, Toronto: Rowman and Allanheld.

Michelson, W. (1988) "Divergent convergence: The daily routines of employed spouses as a public affairs agenda," in C. Andrew and B. Moore Milroy (eds.) *Life Spaces: Gender, Household, Employment*, Vancouver: University of British Columbia Press.

Miller, A. (1990) *The Day Care Dilemma: Critical Concerns for American Families*, New York: Insight Books, Plenum Press.

Minor Matters (A Newsletter about Federal Government Involvement in Child Care), No. 1, 1991. Canberra: Commonwealth Department of Community Services and Health.

Mishra, R., Laws, G. and Hardin, P. (1988) "Ontario," in J. Ismael and Y. Vaillancourt (eds.) *Privatization and Provincial Social Services in Canada: Policy, Administration and Service Delivery*, Edmonton: University of Alberta Press.

Moss, P. (1991) "Day care for young children in the United Kingdom," in E. Melhuish and P. Moss (eds.) *Day Care for Young Children: International Perspectives*, London: Routledge.

Moss, P. and Melhuish, E. C. (1991) "Introduction," in E. C. Melhuish and P. Moss (eds.) *Day Care for Young Children: International Perspectives*, London and New York: Routledge.

National Association of Community Based Children's Services (Australia) (1989) "Funding for childcare," *Policy Issues Forum* April 15–17.

National Center for Education Statistics, US Department of Education (1993) *Digest of Education Statistics 1993*, Washington, D.C.: US Government Printing Office.

Nelson, M. (1991) "The regulation controversy in family day care: The perspective of providers," in J. Shibley Hyde and M. Essex (eds.) *Parental Leave and Child Care*, Philadelphia: Temple University Press.

Nichols, P. (1994) "Los Angeles without traffic jams? It could happen," *American Demographics* 16(2): 18–19.

North, R. N. and Hardwick, W. G. (1992) "Vancouver since the Second World War: An economic geography," in G. Wynn and T. Oke (eds.) *Vancouver and Its Region*, Vancouver: U.B.C. Press.

O'Connell, M. and Bloom, D. E. (1987) *Juggling Jobs and Babies: America's Child Care Challenge* Occasional Papers No. 12, Population Trends and Public Policy, Washington, D.C.: Population Reference Bureau.

Ochiltree, G. (1991) "Child care: A contrast in policies," *Family Matters* (Journal of the Australian Institute of Family Studies), 30: 38–42.

Ontario Advisory Committee on Day Care (1976) *Final Report*, Toronto: Ontario MCSS.

Ontario Coalition for Better Day Care (1987) *Development of Non-Profit Child Care in Ontario*, Brief to the Select Committee on Health, Toronto: Ontario Coalition for Better Day Care.

Ontario Committee on Government Productivity (1973) *Report Number Ten: A Summary*, Toronto.

Ontario Ministry of Community and Social Services (1981) *Day Care Policy: A Background Paper*, Toronto: Program Policy Division, Ontario Ministry of Social and Community Services.

Ontario Ministry of Community and Social Services (1984) *Child Care Services in Ontario: A Background Paper*, Toronto.

Ontario Ministry of Community and Social Services (1987) *New Directions for Child Care*, Toronto: Queen's Printer for Ontario.

Ontario Ministry of Community and Social Services (1991) *Factors Related to Quality in Child Care: A Review of the Literature*, Toronto: Queen's Printer for Ontario.

Ontario Ministry of Community and Social Services (1992a) *Child Care Reform in Ontario: Setting the Stage*, A Public Consultation Paper, Toronto: Queen's Printer for Ontario.

Ontario Ministry of Community and Social Services (1992b) *Ontario Child Care Management Framework*, Toronto: Queen's Printer for Ontario.

Ontario Ministry of Community and Social Services (1993a) *Turning Point: New Support Programs for People with Low-incomes*, Toronto: Queen's Printer for Ontario.

Ontario Ministry of Community and Social Services (1993b) *Study of Non-Profit Child Care Boards in Ontario*, Toronto: Queen's Printer for Ontario.

Ontario Ministry of Community and Social Services (1993c) Special Data Runs (Child Care Branch).

Ontario Ministry of Treasury and Economics (1978) *Ontario Statistics*, Toronto: Queen's Printer of Ontario.

Ontario Select Committee on Health (1987) *Special Report: Future Directions for Child Care in Ontario*, Toronto: Legislative Assembly.

Ontario Social Assistance Review Committee (1988) *Transitions*, Toronto: Queen's Printer for Ontario.

Ontario Task Force on Community and Social Services (1974) *Report on Ministry Goals and Objectives*, Toronto.

Palm, R. and Pred, A. (1978) "The status of women: A time-geographic view," in D. Lanegran and R. Palm (eds.) *An Invitation to Geography*, New York: McGraw-Hill.

Peck, J. T., McCaig, G. and Sapp, M. E. (1988) *Kindergarten Policies: What is Best for Young Children?* Washington, D.C.: National Association for the Education of Young Children.

Phillips, D. (1991) "Day care for young children in the United States," in E. C. Melhuish and P. Moss (eds.) *Day Care for Young Children: International Perspectives*, London and New York: Routledge.

Phillips, D., Howes, C. and Whitebook, M. (1991) "Child care as an adult work environment," *Journal of Social Issues* 47(2): 49–70.

Pickup, L. (1984) "Women's gender-role and its influence on their travel behaviour," *Built Environment* 10: 61–8.

Pickup, L. (1988) "Hard to get around: A study of women's travel mobility," in J. Little, L. Peake and P. Richardson (eds.) *Women in Cities: Gender and the Urban Environment*, New York: New York University Press.

Pickup, L. (1989) "Women's travel requirements: Employment, with domestic constraints," in M. Brieco, L. Pickup and R. Whipp (eds.) *Gender, Transport and Employment: The Impact of Travel Constraints*, Brookfield, Vermont: Gower.

Pinch, S. (1985) *Cities and Services*, London: Routledge and Kegan Paul.

Pinch, S. (1987) "The changing geography of pre-school services in England between 1977 and 1983," *Environment and Planning C* 5: 469–80.

Pinch, S. (1989) "Collective consumption," in J. Wolch and M. Dear (eds.) *The Power of Geography*, Boston: Unwin Hyman.

Pratt, G. and Hanson, S. (1991a) "Geography and the construction of difference," *Gender, Place and Culture* 1(1): 5–29

Pratt, G. and Hanson, S. (1991b) "On the links between home and work: Family-household strategies in a buoyant labour market," *International Journal of Urban and Regional Research* 15: 55–74.

Pratt, G. and Hanson, S. (1993) "Women and work across the life course: Moving beyond essentialism," in C. Katz and J. Monk (eds.) *Full Circles: Geographies of Women over the Life Course*, London and New York: Routledge.

Prentice, S. (1988) "The 'mainstreaming' of daycare," *Resources for Feminist Research/Documentation sur la recherche féministe* 17(3): 59–63.

Press, F. (1991) "The child in children's services: Infrastructure for social justice," *Rattler* 19: 4–7.

Preston, V., McLafferty, S. and Hamilton, E. (1993) "The impact of family status on black, white, and Hispanic women's commuting," *Urban Geography* 14(3): 228–50.

Puleo, V. T. (1988) "A review and critique of research on full-day kindergarten," *The Elementary School Journal* 88: 427–39.

Pupo, N. (1991) "Preserving patriarchy: Women, the family and the state," *Reconstructing the Canadian Family: Feminist Perspectives*, Toronto and Vancouver: Butterworths.

Raneberg, D. and Daubney, T. (1991) "The forgotten consumer: The distorted delivery of child-care services," *Policy* 7(3): 30–3.

Reed, T. (1994) Personal communication (Virginia Tech Resource and Referral Service).

Reeves, D. L. (1992) *Child Care Crisis: A Reference Handbook*, Santa Barbara, California.: ABC-CLIO.

Reinharz, S. (1992) *Feminist Methods in Social Research*, New York: Oxford University Press.

Research Division, National Education Association (1962) *Kindergarten Practices, 1961*, Research Monograph 1962–M2, Washington, D.C.: National Education Association.

Richardson, D. (1993) *Women, Mothering and Childrearing*, London: Macmillan.

Roberts, E. A. M. (1986) "Women's strategies, 1890–1940," in J. Lewis (ed.) *Labour and Love: Women's Experience of Home and Family*, Oxford: Blackwell.

Robins, P. K. and Spiegelman, R. G. (1978) "Substitution among child care modes and the effects of a child care subsidy program," in P. K. Robins and S. Weiner (eds.) *Child Care and Public Policy*, Lexington, Massachusetts: Lexington Books.

Rose, D. (1990) "'Collective consumption' revisited: Analyzing modes of provision and access to child care services in Montréal, Québec," *Political Geography Quarterly* 9(4): 353–80.

Rose, D. (1991) "Access to school daycare services: Class, family, ethnicity and space in Montréal's old and new inner city," *Geoforum* 22(2): 185–201.

Rose, D. (1993) "Local childcare strategies in Montréal, Québec: The mediations of state policies, class and ethnicity in the life courses of families with young children," in C. Katz and J. Monk (eds.) *Full Circles: Geographies of Women over the Life Course*, London and New York: Routledge.

Rose, D. and Chicoine, N. (1991) "Access to school daycare services: Class, family, ethnicity and space in Montréal's old and new inner city," *Geoforum* 22(2): 185–201.

Rose, G. (1993) *Feminism and Geography: The Limits of Geographical Knowledge*, Minneapolis: University of Minnesota Press.

Rosenbloom, S. (1988) "Is there a women's transportation problem?" *Women and Environments* 10: 16–17.

Rosenbloom, S. (1993) "Women's travel patterns at various stages of their lives," in C. Katz and J. Monk (eds.) *Full Circles: Geographies of Women over the Life Course*, London and New York: Routledge.

Royal Commission on the Economic Union and Development Prospects for Canada (1985) *Report*, Ottawa: Ministry of Supply and Services Canada.

Ruchel, J. (1990) "Child care – the challenges for the 1990s," *Rattler* 15: 4–5.

Ruddick, S. (1989) *Maternal Thinking: Towards the Politics of Peace*, New York: Ballantine Books.

Rudolph, M. and Cohen, D. H. (1984) *Kindergarten and Early Schooling*, 2nd edn, Englewood Cliffs: Prentice-Hall Inc.

Rutherford, B. M. and Wekerle, G. R. (1988) "Captive rider, captive labor: Spatial constraints and women's employment," *Urban Geography* 9: 173–93.

Rutherford, B. M. and Wekerle, G. R. (1989) "Single parents in the suburbs: Journey-to-work and access to transportation," *Specialized Transportation Planning and Practice* 3: 277–93.

Salamon, L. (1987) "Of market failure and third-party government: Toward a theory of government–nonprofit relations," in S.A. Ostrander, S. Langton and J. Van Til (eds.) *Shifting the Debate: Public/Private Sector Relations in the Modern Welfare State*, New Brunswick, NJ: Transaction Press.

Sassower, C. A. (1982) "Kindergarten finally becomes universal," *The New York Times* January 10: 41.

Sayer, A. (1992) *Method in Social Science: A Realist Approach*, London: Hutchinson.

Schulz, P. (1978) "Day care in Canada: 1850 to 1962," in K.G. Ross (ed.) *Good Day Care*, Toronto: Women's Press.

Singell, L. D. and Lillydahl, J. (1986) "An empirical analysis of the commute-to-work patterns of males and females in two-earner households," *Urban Studies* 2: 119–29.

Smith, A. and Swain, D. (1988) *Childcare in New Zealand*, Wellington: Allen and Unwin/Port Nicholson Press.

Smith, D. (1977) *Human Geography: A Welfare Approach*, London: Edward Arnold.

Smith, D. (1994) *Geography and Social Justice*, Oxford, UK and Cambridge, MA.: Blackwell.

Smith, D. E. (1986) "Institutional ethnography: A feminist method," *Resources for Feminist Research* 15: 6–13.

Social Planning Council of Metropolitan Toronto (1984) *Caring for Profit: The Commercialization of Human Services in Ontario*, Toronto: Social Planning Council of Metropolitan Toronto.

Sosin, M. (1990) "Decentralizing the social service system: A reassessment," *Social Service Review* 64: 617–36.

Spakes, P. (1991) "A feminist approach to national family policy," in E. Anderson and

R. Hula (eds.) *The Reconstruction of Family Policy*, Westport, Connecticut: Greenwood Press.

Spakes, P. (1992) "National family policy: Sweden versus the United States," *Affilia* 7(2): 44–60.

Stapleford, E. M. (1976) *History of the Day Nurseries Branch*, Toronto: Ontario Ministry of Community and Social Services.

Starrels, M. (1992) "The evolution of workplace family policy research," *Journal of Family Issues* 13(3): 259–78.

State Department of Education (Division of Research, Evaluation and Assessment) (1992) *Town and School District Profiles, 1990–91*, Hartford, Connecticut: State Department of Education.

Statistics Canada (1976) *Census of Canada, 1971 (Vol 1, Part 2): Population, General Characteristics*, Catalogue 92–713, Bulletin 1.2–1. Ottawa: Information Canada.

Statistics Canada (1981) *Census of Canada, 1981: Population, Occcupied Dwellings, Census Families in Private Dwellings: Selected Characteristics, Ontario*, Catalogue 93–918. Ottawa: Information Canada.

Statistics Canada (1984) *Women in the Work World*, Catalogue 99–940. Ottawa: Ministry of Supply and Services.

Statistics Canada (1992) *Census of Canada 1991: Profiles of Census Divisions and Subdivisions in Ontario – Part A*, Catalogue 95–337. Ottawa: Ministry of Supply and Services.

Statistics Canada (monthly) *Labour Force Survey*, Catalogue 71–001. Ottawa: Ministry of Supply and Services.

Steiner, A. K. (1957) "A report on state laws: Early elementary education," *School Life* 39: 7–10.

Stolzenberg, R. M. and Waite, L. J. (1984) "Local labor markets, children, and labor force participation of wives," *Demography* 21: 157–70.

Sweeney, T. (1989) "Inequalities in our provisions for young children," in R. Kennedy (ed.) *Australian Welfare: Historical Sociology*, Melbourne: Macmillan.

Thorman, G. (1989) *Day Care . . . An Emerging Crisis*, Springfield, Illinois: Charles C. Thomas.

Tietjen, A. M. (1985) "The social networks and social support of married and single mothers in Sweden," *Journal of Marriage and the Family* 47: 489–96.

Tivers, J. (1985) *Women Attached: The Daily Lives of Women with Young Children*, London: Croom Helm.

Tivers, J. (1988) "Women with young children: Constraints on activities in the urban environment," in J. Little, L. Peake and P. Richardson (eds.) *Women in Cities: Gender and the Urban Environment*, New York: New York University Press.

Townson, M. (1985) *Financing Child Care through the Canada Assistance Plan*, Ottawa: Task Force on Child Care.

Trost, C. (1988) "School child-care programs stir debate," *The Wall Street Journal* April 8: 17.

Truelove, M. (1989) "Preschool day care and public facility provision: A case study of Metropolitan Toronto," PhD Thesis, Department of Geography, University of Toronto.

Truelove, M, (1993a) "Measurement of spatial equity," *Environment and Planning C: Government and Policy* 11: 19–34.

Truelove, M. (1993b) *Location Decisions for Non-Profit, Commercial and Government Day Care Centres in the Toronto Census Metropolitan Area*, Toronto: School of Applied Geography, Ryerson Polytechnic University.

Tuominen, M. (1991) "Caring for profit: Social, economic, and political significance of for-profit child care," *Social Service Review* September: 450–67.

Ulrey, G. L., Alexander, K., Bender, B. and Gillis, H. (1982) "Effects of length of school day on kindergarten school performance and parent satisfaction," *Psychology in the Schools* 19: 238–42.

US Bureau of Labor Statistics (1990) *Occupational Projections and Training Data*, Washington D.C.: US Government Printing Office.

US Bureau of Labor Statistics (1991) *Employee Benefits in Small Private Establishments, 1990*, Washington, D.C.: US Government Printing Office.

US Bureau of Labor Statistics (1992) *Employee Benefits in State and Local Governments, 1990*, Washington, D.C.: US Government Printing Office.

US Bureau of Labor Statistics (1993) *Employee Benefits in Medium and Large Private Establishments, 1991*, Washington, D.C.: US Government Printing Office.

US Bureau of the Census (1992a) *Statistical Abstract of the United States: 1992*, 112th edn, Washington, D.C.: US Government Printing Office.

US Bureau of the Census (1992b) *1990 Census of Population and Housing Summary Tapefile 3A Connecticut*, Washington, D.C.: US Department of Commerce.

US Bureau of the Census (1992c) *1990 Census of Population and Housing Summary Tapefile 3C Connecticut*, Washington, D.C.: US Department of Commerce.

US Bureau of the Census (1994a) *Statistical Abstract of the United States, 1994*, Washington, D.C.: US Government Printing Office.

US Bureau of the Census (1994b) *Who's Minding the Kids? Child Care Arrangements: Fall 1991*, Washington, D.C.: US Government Printing Office.

Walsh, D. J. (1989) "Changes in kindergarten: Why here? Why now?," *Early Childhood Research Quarterly* 4: 377–91.

Warren, R., Rosentraub, M. S. and Weschler, L. F. (1988) "A community services budget: Public, private, and third-sector roles in urban services," *Urban Affairs Quarterly* 23: 414–31.

Wattenberg, B. (1987) *The Birth Dearth*, New York: Pharos.

Weber, E. (1969) *The Kindergarten: Its Encounter with Education Thought in America*, New York: Teachers College Press, Columbia University.

Weiner, S. (1978) "The child care market in Seattle and Denver," in P. K. Robins and S. Weiner (eds.) *Child Care and Public Policy*, Lexington, MA: Lexington Books.

Wekerle, G. R. (1980) "Women in the urban environment," *Signs* 5: 188–213.

Wekerle, G. and Rutherford, B. (1989) "The mobility of capital and the immobility of female labor: Responses to economic restructuring," in J. Wolch and M. Dear (eds.) *The Power of Geography*, London: Allen and Unwin.

Wharf, B. and Cossom, J. (1987) "Citizen participation and social welfare policy," in S. Yelaja (ed.) *Canadian Social Policy*, Waterloo: Wilfrid Laurier Press.

Williams, F. (1989) *Social Policy: A Critical Introduction. Issues of Race, Gender and Class*, Oxford: Blackwell.

Wolch, J. (1990) *The Shadow State: Government and Voluntary Sector in Transition*, New York: The Foundation Center.

Wolcott, I. (1991) "Work and family: Employers' views," *Australian Institute of Family Studies Monograph No. 11*, Melbourne.

Women and Environments (1988) "California going for child care in transit," *Women and Environments* 12: 19.

Women's Bureau (1981) *Workplace Child Care: A Background Paper*, Toronto: Ontario Ministry of Labour.

Youcha, G. (1995) *Minding the Children*, New York: Scribner (Simon and Schuster).

Young, A. H. and Good, C. A. (1986) "Working mothers and child care in New Orleans," *Urban Resources* 3(2): 29–32.

Zigler, E. F. and Gordon, E. W. (eds.) (1982) *Day Care: Scientific and Social Policy Issues*, Boston: Auburn House.

Zigler, E. F. and Lang, M. E. (1991) *Child Care Choices: Balancing the Needs of Children, Families, and Society*, New York: The Free Press (Macmillan).

INDEX

Act for Better Child Care Services (1988) 26–7
active agents, women as 15–16, 91, 112, 123, 139
activism, women's 145, 173, 176–7; and human agency 170; and the state 157–61
after-school child care 136
aged care 166
agency, human: and women's activism 170
Aid to Families with Dependent Children (AFDC) 10, 12
At Risk program: US 27
Australia: child care 5, 154, 155, 157, 162; for-profit child care 163–4; subsidies 147, 148–9, 164; welfare systems 151; women's activism 159–60, 160–1
Austria 150

Barnard, Henry 52
benefits: family 8; from employers 9, 32–4, 110, 120
Blacksburg, Virginia 15, 173, 175; child care centers 82–92
Boston 52
Britain see UK
British Columbia 41

Canada: child care 4–5, 156, 172; child care workers 152–3; for-profit child care 162; government policies 3, 6–13, 36–45, 147, 154, 169; maternity leave 174; non-profit child care 12, 44, 64–5, 95–107, 175–6; women's activism 160; see also Metropolitan Toronto; Ontario; Vancouver
Canada Assistance Plan (CAP) 10, 12, 45,

66; funding for child care 40–2
cash rebates 160
chains: child care 7, 148, 162
Child and Dependent Care Tax Credit 13, 25–6
child care: everyday lives of working mothers 109–22; full-day kindergartens 49–61; international policies 143–66; issues 3–6, 169–77; journeys to 77–92; local support networks 123–40; see also Canada; US
Child Care and Development Block Grants 10, 27
child care centers: Canada 38–45; funding 6–7; and income 11, 154; journeys to 77–92; Metropolitan Toronto 93–108; Ontario 62–74; US 28, 31, 110; see also for-profit; non-profit; workplace
child care chains 7, 148, 162
Child Care Expense Deduction: Canada 41
child care industry 24–5, 30–2, 162, 174
Child Care Reform: Canada 68, 175
child care workers: Canada 44, 45; ethnicity 172; home-based 12–13; journeys to child care 88–9; nannies 6, 12–13, 137; pay 31, 44, 149, 152–3, 163, 172; relationship with children 134; in underground economy 12–13, 26; US 24, 31
child development: and regulation 164
children: effects of child care 34–5, 110, 118–19; effects of journeys to child care 78, 86; environment issue 10, 52, 57–8, 59–60; help from older 115–16, 121–2; maintenance of social relationships 135, 136; relationship

200